TEACHER RESOURCES

Diversity of Life

Full Option Science System
Developed at the Lawrence Hall of Science, University of California, Berkeley
Published and Distributed by Delta Education

FOSS Lawrence Hall of Science Team
Larry Malone and Linda De Lucchi, FOSS Project Codirectors and Lead Developers
Kathy Long, FOSS Assessment Director; David Lippman, Program Manager; Carol Sevilla, Publications Design Coordinator; Susan Stanley, Illustrator; John Quick, Photographer
FOSS Curriculum Developers: Brian Campbell, Teri Lawson, Alan Gould, Susan Kaschner Jagoda, Ann Moriarty, Jessica Penchos, Kimi Hosoume, Virginia Reid, Joanna Snyder, Erica Beck Spencer, Joanna Totino, Diana Velez, Natalie Yakushiji
Susan Ketchner, Technology Project Manager
FOSS Technology Team: Dan Bluestein, Christopher Cianciarulo, Matthew Jacoby, Kate Jordan, Frank Kusiak, Nicole Medina, Jonathan Segal, Dave Stapley, Shan Tsai

Delta Education Team
Bonnie A. Piotrowski, Editorial Director, Elementary Science
Project Team: Jennifer Apt, Sandra Burke, Mary Connell, Joann Hoy, Angela Miccinello, Jennifer Staltare

Content Reviewer
Gillian Puttick, TERC

Thank you to all FOSS Middle School Revision Trial Teachers and District Coordinators
Frances Amojioyi, Lincoln Middle School, Alameda, CA; Dean Anderson, Organized trials for Boston Public Schools, Boston, MA; Thomas Archer, Organized trials for ESD 112, Vancouver, WA; Lauresa Baker, Lincoln Middle School, Alameda, CA; Bobbi Anne Barnowsky, Canyon Middle School, Castro Valley, CA; Christine Bertko, St. Finn Barr Catholic School, San Francisco, CA; Stephanie Billinge, James P. Timilty Middle School, Roxbury, MA; Jerry Breton, Ingleside Middle School, Phoenix, AZ; Robert Cho, Timilty Middle School, Boston, MA; Susan Cohen, Cherokee Heights Middle School, Madison, WI; Malcolm Davis, Canyon Middle School, Castro Valley, CA; Marilyn Decker, Organized trials for Milton PS, Milton, MA; Jenny Ernst, Park Day School, Oakland, CA; Marianne Floyd, Spanaway Middle School, Spanaway, WA; Sarah Kathryn Gessford, Journeys School, Jackson, WY; Charles Hardin, Prairie Point Middle School, Cedar Rapids, IA; Jennifer Hartigan, Lincoln Middle School, Alameda, CA; Sheila Holland, TechBoston Academy, Boston, MA; Nicole Hoyceanyls, Charles S. Pierce Middle School, Milton, MA; Bruce Kamerer, Donald McKay K-8 School, East Boston, MA; Carmen Saele Kardokus, Reeves Middle School, Olympia, WA; Janey Kaufman, Organized trials for Scottsdale USD, Scottsdale, AZ; Erica Larson, Organized trials for Grant Wood AEA, Cedar Rapids, IA; Lindsay Lodholz, O'Keeffe Middle School, Madison, WI; Robert Mattisinko, Chaparral High School, Scottsdale, AZ; Brenda McGurk, Prairie Point Middle School, Cedar Rapids, IA; Tim Miller, Mountainside Middle School, Scottsdale, AZ; Thomas Miro, Lincoln Middle School, Alameda, CA; Spencer Nedved, Frontier Middle School, Vancouver, WA; Joslyn Olsen, Lincoln Middle School, Alameda, CA; Stephanie Ovechka, Cedarcrest Middle School, Spanaway, WA; Barbara Reinert, Copper Ridge School, Scottsdale, AZ; Stephen Ramos, Lincoln Middle School, Alameda, CA; Gina Rutenbeck, Prairie Point Middle School, Cedar Rapids, IA; John Sheridan, Boston Public Schools (Boston Schoolyard Initiative), Boston, MA; Barbara Simon, Timilty Middle School, Boston, MA; Lise Simpson, Alcott Middle School, Norman, OK; Autumn Stevick, Thurgood Marshall Middle School, Olympia, WA; Ted Stoeckley, Hall Middle School, Larkspur, CA; Lesli Taschwer, Organized trials for Madison SD, Madison, WI; Paula Warner, Alcott Middle School, Norman, OK; Darren T. Wells, James P. Timilty Middle School, Boston, MA; Kristin White, Frontier Middle School, Vancouver, WA

Photo Credits: © hans engbers/Shutterstock (cover); © Delta Education

Published and Distributed by Delta Education, a member of the School Specialty Family
The FOSS program was developed in part with the support of the National Science Foundation grant nos. ESI-9553600 and ESI-0242510. However, any opinions, findings, conclusions, statements, and recommendations expressed herein are those of the authors and do not necessarily reflect the views of NSF. FOSSmap was developed in collaboration between the BEAR Center at UC Berkeley and FOSS at the Lawrence Hall of Science.

Copyright © 2018 by The Regents of the University of California

All rights reserved. Any part of this work (other than duplication masters) may not be reproduced or transmitted in any form or by any means, electronic or mechanical, including photocopying and recording, or by an information storage or retrieval system without prior written permission. For permission please write to: FOSS Project, Lawrence Hall of Science, University of California, Berkeley, CA 94720 or foss@berkeley.edu.

Diversity of Life — Teacher Toolkit, 1558524
Teacher Resources, 1558539
978-1-62571-808-2
Printing 2 – 6/2017
Patterson Printing, Benton Harbor, MI

WARNING — This set contains chemicals that may be harmful if misused. Read cautions on individual containers carefully. Not to be used by children except under adult supervision.

This warning label is required by the
U.S. Consumer Product Safety Commission.
The chemicals in the FOSS Diversity of Life kit are
camphor, congo red stain, polyacrylate crystals, kosher salt,
and methylene blue stain.

TEACHER RESOURCES

TABLE OF CONTENTS

FOSS Program Goals **A1**

Science Notebooks in Middle School **B1**

Science-Centered Language Development in Middle School **C1**

FOSSweb and Technology **D1**

Science Notebook Masters **1–67**

Teacher Masters .... **A–DDD**

Assessment Masters
Assessment Charts 1–25
Entry-Level Survey 1–4
Investigations 1–3 I-Check 1–4
Investigation 4 I-Check 1–4
Investigation 5 I-Check 1–4
Investigation 6 I-Check 1–4
Investigation 7 I-Check 1–4
Posttest 1–7

Notebook Answers

This document, *Teacher Resources*, is one of three parts of the *FOSS Teacher Toolkit* for this course. The chapters in *Teacher Resources* are all available as PDFs on FOSSweb.

The other parts of the course *Teacher Toolkit* are the *Investigations Guide* and a copy of the *FOSS Science Resources* student book containing original readings for this course.

The spiral-bound *Investigations Guide* contains these chapters.

- Overview
- Framework and NGSS
- Materials
- Investigations
- Assessment

The *Teacher Toolkit* is the most important part of the FOSS Program. It is here that all the wisdom and experience contributed by hundreds of educators has been assembled. Everything we know about the content of the course, how to teach the subject, and the resources that will assist the effort are presented here.

FOSS Program Goals

FOSS Program Goals

Contents

Introduction **A1**

Goals of the FOSS Program ... **A2**

Bridging Research
into Practice **A5**

FOSS Next Generation
K–8 Scope and Sequence **A8**

INTRODUCTION

The Full Option Science System™ has evolved from a philosophy of teaching and learning at the Lawrence Hall of Science that has guided the development of successful active-learning science curricula for more than 40 years. The FOSS Program bridges research and practice by providing tools and strategies to engage students and teachers in enduring experiences that lead to deeper understanding of the natural and designed worlds.

Science is a creative and analytic enterprise, made active by our human capacity to think. Scientific knowledge advances when scientists observe objects and events, think about how they relate to what is known, test their ideas in logical ways, and generate explanations that integrate the new information into understanding of the natural world. Engineers apply that understanding to solve real-world problems. Thus the scientific enterprise is both what we know (content knowledge) and how we come to know it (practices). Science is a discovery activity, a process for producing new knowledge.

The best way for students to appreciate the scientific enterprise, learn important scientific and engineering concepts, and develop the ability to think well is to actively participate in scientific practices through their own investigations and analyses. FOSS was created to engage students and teachers with meaningful experiences in the natural and designed worlds.

 Full Option Science System

FOSS Program Goals

GOALS OF THE FOSS PROGRAM

FOSS has set out to achieve three important goals: scientific literacy, instructional efficiency, and systemic reform.

Scientific Literacy

FOSS provides all students with science experiences that are appropriate to students' cognitive development and prior experiences. It provides a foundation for more advanced understanding of core science ideas that are organized in thoughtfully designed learning progressions and prepares students for life in an increasingly complex scientific and technological world.

The National Research Council (NRC) in *A Framework for K–12 Science Education: Practices, Crosscutting Concepts, and Core Ideas* and the American Association for the Advancement of Science (AAAS) in *Benchmarks for Scientific Literacy* have described the characteristics of scientific literacy:

- Familiarity with the natural world, its diversity, and its interdependence.

- Understanding the disciplinary core ideas and the crosscutting concepts of science, such as patterns; cause and effect; scale, proportion, and quantity; systems and system models; energy and matter—flows, cycles, and conservation; structure and function; and stability and change.

- Knowing that science and engineering, technology, and mathematics are interdependent human enterprises and, as such, have implied strengths and limitations.

- Ability to reason scientifically.

- Using scientific knowledge and scientific and engineering practices for personal and social purposes.

The FOSS Program design is based on learning progressions that provide students with opportunities to investigate core ideas in science in increasingly complex ways over time. FOSS starts with the intuitive ideas that primary students bring with them and provides experiences that allow students to develop more sophisticated understanding as they grow through the grades. Cognitive research tells us that learning involves individuals in actively constructing schemata to organize new information and to relate and incorporate the new understanding into established knowledge. What sets experts apart from novices is that experts in a discipline have extensive knowledge that is effectively organized into structured schemata to promote thinking. Novices have disconnected ideas about a topic that are difficult to retrieve and use. Through internal processes to establish schemata and through social processes of interacting with peers and adults, students construct understanding of the natural world and their relationship to it.

The target goal for FOSS students is to know and use scientific explanations of the natural world and the designed world; to understand the nature and development of scientific knowledge and technological capabilities; and to participate productively in scientific and engineering practices.

Instructional Efficiency

FOSS provides all teachers with a complete, cohesive, flexible, easy-to-use science program that reflects current research on teaching and learning, including student discourse, argumentation, writing to learn, and reflective thinking, as well as teacher use of formative assessment to guide instruction. The FOSS Program uses effective instructional methodologies, including active learning, scientific practices, focus questions to guide inquiry, working in collaborative groups, multisensory strategies, integration of literacy, appropriate use of digital technologies, and making connections to students' lives.

FOSS is designed to make active learning in science engaging for teachers as well as for students. It includes these supports for teachers:

- Complete equipment kits with durable, well-designed materials for all students.
- Detailed *Investigations Guide* with science background for the teacher and focus questions to guide instructional practice and student thinking.
- Multiple strategies for formative assessment.
- Benchmark assessments with scoring guides.
- Strategies for use of science notebooks for novice and experienced users.
- *FOSS Science Resources*, a book of course-specific readings.
- The FOSS website with course-integrated online activities for use in school or at home, suggested extension activities, and extensive online support for teachers.

FOSS Program Goals

Systemic Reform

FOSS provides schools and school systems with a program that addresses the community science-achievement standards. The FOSS Program prepares students by helping them acquire the knowledge and thinking capacity appropriate for world citizens.

The FOSS Program design makes it appropriate for reform efforts on all scales. It reflects the core ideas to be incorporated into the next-generation science standards. It meets with the approval of science and technology companies working in collaboration with school systems, and it has demonstrated its effectiveness with diverse student and teacher populations in major urban reform efforts. The use of science notebooks and formative-assessment strategies in FOSS redefines the role of science in a school—the way that teachers engage in science teaching with one another as professionals and with students as learners, and the way that students engage in science learning with the teacher and with one another. FOSS takes students and teachers beyond the classroom walls to establish larger communities of learners.

BRIDGING RESEARCH INTO PRACTICE

The FOSS Program is built on the assumptions that understanding core scientific knowledge and how science functions is essential for citizenship, that all teachers can teach science, and that all students can learn science. The guiding principles of the FOSS design, described below, are derived from research and confirmed through FOSS developers' extensive experience with teachers and students in typical American classrooms.

Understanding of science develops over time. FOSS has elaborated learning or content progressions for core ideas in science for kindergarten through grade 8. Developing the learning progressions involves identifying successively more sophisticated ways of thinking about core ideas over multiple years. "If mastery of a core idea in a science discipline is the ultimate educational destination, then well-designed learning progressions provide a map of the routes that can be taken to reach that destination" (National Research Council, *A Framework for K–12 Science Education*, 2011).

Focusing on a limited number of topics in science avoids shallow coverage and provides more time to explore core science ideas in depth. Research emphasizes that fewer topics experienced in greater depth produces much better learning than many topics briefly visited. FOSS affirms this research. FOSS courses provide long-term engagement (10–12 weeks) with important science ideas. Furthermore, courses build upon one another within and across each strand, progressively moving students toward the grand ideas of science. The core ideas of science are difficult and complex, never learned in one lesson or in one class year.

FOSS Next Generation—K–8 Sequence

		PHYSICAL SCIENCE		EARTH SCIENCE		LIFE SCIENCE	
		MATTER	ENERGY AND CHANGE	ATMOSPHERE AND EARTH	ROCKS AND LANDFORMS	STRUCTURE/ FUNCTION	COMPLEX SYSTEMS
8	6–8	Waves; Gravity and Kinetic Energy; Chemical Interactions; Electromagnetic Force; Variables and Design		Planetary Science; Earth History; Weather and Water		Heredity and Adaptation; Human Systems Interactions; Populations and Ecosystems; Diversity of Life	
	5	Mixtures and Solutions		Earth and Sun		Living Systems	
	4		Energy		Soils, Rocks, and Landforms	Environments	
	3	Motion and Matter		Water and Climate		Structures of Life	
	2	Solids and Liquids			Pebbles, Sand, and Silt	Insects and Plants	
	1		Sound and Light	Air and Weather		Plants and Animals	
K	K	Materials and Motion		Trees and Weather		Animals Two by Two	

FOSS Program Goals

Science is more than a body of knowledge. How well you think is often more important than how much you know. In addition to the science content framework, every FOSS course provides opportunities for students to engage in and understand science practices, and many courses explore issues related to engineering practices and the use of natural resources. FOSS uses these science and engineering practices.

- Asking questions (for science) and defining problems (for engineering)
- Developing and using models
- Planning and carrying out investigations
- Analyzing and interpreting data
- Using mathematics, information and computer technology, and computational thinking
- Constructing explanations (for science) and designing solutions (for engineering)
- Engaging in argument from evidence
- Obtaining, evaluating, and communicating information

Science is inherently interesting, and children are natural investigators. It is widely accepted that children learn science concepts best by doing science. Doing science means hands-on experiences with objects, organisms, and systems. Hands-on activities are motivating for students, and they stimulate inquiry and curiosity. For these reasons, FOSS is committed to providing the best possible materials and the most effective procedures for deeply engaging students with scientific concepts. FOSS students at all grade levels investigate, experiment, gather data, organize results, and draw conclusions based on their own actions. The information gathered in such activities enhances the development of science and engineering practices.

Education is an adventure in self-discovery. Science provides the opportunity to connect to students' interests and experiences. Prior experiences and individual learning styles are important considerations for developing understanding. Observing is often equated with seeing, but in the FOSS Program all senses are used to promote greater understanding. FOSS evolved from pioneering work done in the 1970s with students with disabilities. The legacy of that work is that FOSS investigations naturally use multisensory methods to accommodate students with physical and learning disabilities and also to maximize information gathering for all students. A number of tools, such as the FOSS syringe and balance, were originally designed to serve the needs of students with disabilities.

Formative assessment is a powerful tool to promote learning and can change the culture of the learning environment. Formative assessment in FOSS creates a community of reflective practice. Teachers and students make up the community and establish norms of mutual support, trust, respect, and collaboration. The goal of the community is that everyone will demonstrate progress and will learn and grow.

Science-centered language development promotes learning in all areas. Effective use of science notebooks can promote reflective thinking and contribute to lifelong learning. Research has shown that when language-arts experiences are embedded within the context of learning science, students improve in their ability to use their language skills. Students are motivated to read to find out information, and to share their experiences both verbally and in writing.

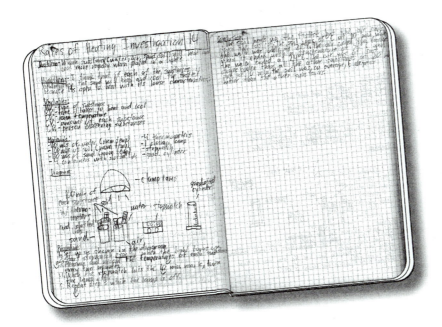

Experiences out of the classroom develop awareness of community. By extending classroom learning into the local region and community, FOSS brings the science concepts and principles to life. In the process of extending classroom learning to the natural world and utilizing community resources, students will develop a relationship with learning that extends beyond the classroom walls.

FOSS Program Goals

FOSS Program Goals

FOSS NEXT GENERATION K-8 SCOPE AND SEQUENCE

Grade	Physical Science	Earth Science	Life Science
6–8	Waves* Gravity and Kinetic Energy*	Planetary Science	Heredity and Adaptation* Human Systems Interactions*
	Chemical Interactions	Earth History	Populations and Ecosystems
	Electromagnetic Force* Variables and Design*	Weather and Water	Diversity of Life
5	Mixtures and Solutions	Earth and Sun	Living Systems
4	Energy	Soils, Rocks, and Landforms	Environments
3	Motion and Matter	Water and Climate	Structures of Life
2	Solids and Liquids	Pebbles, Sand, and Silt	Insects and Plants
1	Sound and Light	Air and Weather	Plants and Animals
K	Materials and Motion	Trees and Weather	Animals Two by Two

* Half-length course

FOSS is a research-based science curriculum for grades K–8 developed at the Lawrence Hall of Science, University of California, Berkeley. FOSS is also an ongoing research project dedicated to improving the learning and teaching of science. The FOSS project began over 25 years ago during a time of growing concern that our nation was not providing young students with an adequate science education. The FOSS Program materials are designed to meet the challenge of providing meaningful science education for all students in diverse American classrooms and to prepare them for life in the 21st century. Development of the FOSS Program was, and continues to be, guided by advances in the understanding of how people think and learn.

With the initial support of the National Science Foundation and continued support from the University of California, Berkeley, and School Specialty, Inc., the FOSS Program has evolved into a curriculum for all students and their teachers, grades K–8. The current editions of FOSS are the result of a rich collaboration among the FOSS/Lawrence Hall of Science development staff; the FOSS product development team at School Specialty; assessment specialists, educational researchers, and scientists; and dedicated professionals in the classroom and their students, administrators, and families.

We acknowledge the thousands of FOSS educators who have embraced the notion that science is an active process, and we thank them for their significant contributions to the development and implementation of the FOSS Program.

Science Notebooks in Middle School

Science Notebooks in Middle School

A scientist's notebook

A student's notebook

INTRODUCTION

Scientists keep notebooks. The scientist's notebook is a detailed record of his or her engagement with scientific phenomena. It is a personal representation of experiences, observations, and thinking—an integral part of the process of doing scientific work. A scientist's notebook is a continuously updated history of the development of scientific knowledge and reasoning. The notebook organizes the huge body of knowledge and makes it easier for a scientist to work. As developing scientists, FOSS students are encouraged to incorporate notebooks into their science learning. First and foremost, the notebook is a tool for student learning.

Contents

Introduction	B1
Notebook Benefits	B2
Getting Started	B5
Notebook Components	B12
Focusing the Investigation	B14
Data Acquisition and Organization	B16
Making Sense of Data	B18
Next-Step Strategies	B22
Using Notebooks to Improve Student Learning	B25
Derivative Products	B28

Science Notebooks in Middle School

From the Human Systems Interactions Course

NOTEBOOK BENEFITS

Engaging in active science is one part experience and two parts making sense of the experience. Science notebooks help students with the sense-making part by providing two major benefits: documentation and cognitive engagement.

Benefits to Students

Science notebooks centralize students' data. When data are displayed in functional ways, students can think about the data more effectively. A well-kept notebook is a useful reference document. When students have forgotten a fact or relationship that they learned earlier in their studies, they can look it up. Learning to reference previous discoveries and knowledge structures is important.

Documentation: an organized record. As students become more accomplished at keeping notebooks, their work will become better organized and efficient. Tables, graphs, charts, drawings, and labeled illustrations will become standard means for representing and displaying data. A complete and accurate record of learning allows students to reconstruct the sequence of learning events and relive the experience. Discussions about science among students, students and teachers, or students, teachers, and families, have more meaning when they are supported by authentic documentation in students' notebooks. Questions and ideas generated by experimentation or discussion can be recorded for future investigation.

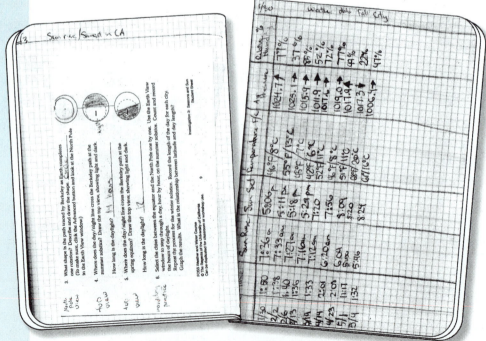

From the Weather and Water Course

Full Option Science System

Cognitive engagement. Once data are recorded and organized in an efficient manner in science notebooks, students can think about the data and draw conclusions about the way the world works. Their data are the raw materials that students use to forge concepts and relationships from their experiences and observations.

Writing stimulates active reasoning. There is a direct relationship between the formation of concepts and the rigors of expressing them in words. Writing requires students to impose discipline on their thoughts. When you ask students to generate derivative products (summary reports, detailed explanations, posters, oral presentations, etc.) as evidence of learning, the process will be much more efficient and meaningful because they have a coherent, detailed notebook for reference.

When students use notebooks as an integral part of their science studies, they think critically about their thinking. This reflective thinking can be encouraged by notebook entries that present opportunities for self-assessment. Self-assessment motivates students to rethink and restate their scientific understanding. Revising their notebook entries helps students clarify their understanding of the science concepts under investigation. By writing explanations, students clarify what they know and expose what they don't know.

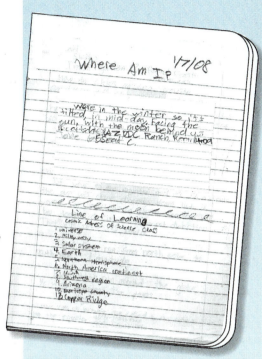

From the Planetary Science Course

From the Planetary Science Course

Science Notebooks in Middle School

Science Notebooks in Middle School

Benefits to Teachers

In FOSS, the unit of instruction is the course—a sequence of conceptually related learning experiences that leads to a set of learning outcomes. A science notebook helps you think about and communicate the conceptual structure of the course you are teaching.

Assessment. From the assessment point of view, a science notebook is a collection of student-generated artifacts that exhibit learning. You can informally assess student skills, such as using charts to record data, in real time while students are working with materials. At other times, you might collect student work samples and review them for insights or errors in conceptual understanding. This valuable information helps you plan the next steps of instruction. Students' data analysis, sense making, and reflection provide a measure of the quality and quantity of student learning. The notebook itself should not be graded, though certain assignments might be graded and placed in the notebook.

Medium for feedback. The science notebook provides an excellent medium for providing feedback to individual students regarding their work. Productive feedback calls for students to read a teacher comment, think about the issue it raises, and act on it. The comment may ask for clarification, an example, additional information, precise vocabulary, or a review of previous work in the notebook. In this way, you can determine whether a problem with the student work relates to a flawed understanding of the science content or a breakdown in communication skills.

Focus for professional discussions. The student notebook also acts as a focal point for discussion about student learning at several levels. First, a student's work can be the subject of a conversation between you and the student. By acting as a critical mentor, you can call attention to ways a student can improve the notebook, and help him or her learn how to use the notebook as a reference. You can also review and discuss the science notebook during family conferences. Science notebooks shared among teachers in a study group or other professional-development environment can effectively demonstrate recording techniques, individual styles, various levels of work quality, and so on. Just as students can learn notebook strategies from one another, teachers can learn notebook skills from one another.

GETTING STARTED

A middle school science notebook is more than just a collection of science work, notes, field-trip permission slips, and all the other types of documents that tend to accumulate in a student's three-ring binder or backpack. By organizing the science work systematically into a bound composition book, students create a thematic record of their experiences, thoughts, plans, reflections, and questions as they work through a topic in science.

The science notebook is more than just formal lab reports; it is a record of a student's entire journey through a progression of science concepts. Where elementary school students typically need additional help structuring and organizing their written work, middle school students should be encouraged to develop their organizational skills and take some ownership in creating deliberate records of their science learning, even though they may still require some pointers and specific scaffolding from you.

In addition, the science notebook provides a personal space where students can explore their understanding of science concepts by writing down ideas and being allowed to "mess around" with their thinking. Students are encouraged to look back on their ideas throughout the course to self-assess their conceptual development and record new thoughts. With this purpose of the science notebook in mind, you may need to refine your own thinking around what should or should not be included as a part of the science notebook, as well as expectations about grading and analyzing student work.

Science Notebooks in Middle School

Rules of Engagement

Teachers and students should be clear about the conventions students will honor in their notebook entries. Typically, the rules of grammar and spelling are fairly relaxed so as not to inhibit the flow of expression during notebook entries. This also helps students develop a sense of ownership in their notebooks, a place where they are free to write in their own style. When students generate derivative products using information in the notebooks, such as reports, you might require students to exercise more rigorous language-arts conventions.

In addition to written entries, students should be encouraged to use a wide range of other means for recording and communicating, including charts, tables, graphs, drawings, graphics, color codes, numbers, and artifacts attached to the notebook pages. By expanding the options for making notebook entries, each student will find his or her most efficient, expressive way to capture and organize information for later retrieval.

Enhanced Classroom Discussion

One of the benefits of using notebooks is that you will elicit responses to key discussion questions from all students, not just the handful of verbally enthusiastic students in the class. When you ask students to write down their thoughts after you pose a question, all students have time to engage deeply with the question and organize their thoughts. When you ask students to share their answers, those who needed more time to process the question and organize their thinking will be ready to verbalize their responses and become involved in a class discussion.

When students can use their notebooks as a reference during the ensuing discussion, they won't feel put on the spot. At some points, you might ask students to share only what they wrote in their notebooks, to remind them to focus their thoughts while writing. As the class shares ideas during discussions, students can add new ideas to their notebooks under a line of learning (see next-step strategies). Even if some students are still reticent, having students write after a question is posed prevents them from automatically disengaging from conversations.

Notebook Structure

FOSS recommends that students keep their notebooks in 8" × 10" bound composition books. At the most advanced level, students are responsible for creating the entire science notebook from blank pages in their composition books. Experienced students determine when to use their notebooks, how to organize space, what methods of documentation to use, and how to flag important information. This level of notebook use will not be realized quickly; it will likely require systematic development by an entire teaching staff over time.

At the beginning, notebook practice is often highly structured, using prepared sheets from the FOSS notebook masters. You can photocopy and distribute these sheets to students as needed during the investigations. Sheets are sized to fit in a standard composition book. Students glue or tape the sheets into their notebooks. This allows some flexibility between glued-in notebook sheets and blank pages where students can do additional writing, drawings, and other documentation. Prepared notebook sheets are helpful organizers for students with challenges such as learning disabilities or with developing English skills. This model is the most efficient means for obtaining the most productive work from inexperienced middle school students.

From the Planetary Science Course

Science Notebooks in Middle School

Science Notebooks in Middle School

To make it easy for new FOSS teachers to implement a beginning student notebook, Delta Education sells copies of the printed *FOSS Science Notebook* in English for all FOSS middle school courses. Electronic versions of the notebook sheets can be downloaded free of charge at www.FOSSweb.com.

Each *FOSS Science Notebook* is a bound set of the notebook sheets for the course plus extra blank sheets throughout the notebook for students to write focus or inquiry questions, record and organize data, make sense of their thinking, and write summaries. There are also blank pages at the end to develop an index of science vocabulary.

The questions, statements, and graphic organizers on the notebook sheets provide guidance for students and scaffolding for teachers. When the notebook sheets are organized as a series, they constitute a highly structured precursor to an autonomously generated science notebook.

Developing Notebook Skills

Students will initially need more guidance from you. You will need to describe what and when to record, and to model organizational techniques. As the year advances, the notebook work will become increasingly student centered. As the body of work in the notebook grows, students will have more and more examples of useful techniques for reference. This self-sufficiency reduces the amount of guidance you need to provide, and reinforces students' appreciation of their own record of learning.

This gradual shift toward student-centered use of the notebook applies to any number of notebook skills, including developing headers for each page (day, time, date, title, etc.); using space efficiently on the page; preparing graphs, graphic organizers, and labeled illustrations; and attaching artifacts (sand samples, dried flowers, photographs, etc.). For instance, when students first display their data in a two-coordinate graph, the graph might be completely set up for them, so that they simply plot the data. As the year progresses, they will be expected to produce graphs with less and less support, until they are doing so without any assistance from you.

Science Notebooks in Middle School

Organizing Science Notebooks

Four organizational components of the notebook should be planned right from the outset: a table of contents, page numbering, entry format, and an index.

Table of contents. Students should reserve the first three to five pages of their notebook for the table of contents. They will add to it systematically as they proceed through the course. The table of contents should include the date, title, and page number for each entry. The title could be based on the names of the investigations in the course, the specific activities undertaken, the concepts learned, a focus question for each investigation, or some other schema that makes sense to everyone.

Page numbering. Each page should have a number. These are referenced in the table of contents as the notebook progresses.

Entry format. During each class session, students will document their learning. Certain information will appear in every record, such as the date and title. Other forms of documentation will vary, including different types of written entries and artifacts, such as a multimedia printout. Some teachers ask their students to start each new entry at the top of the next available page. Others simply leave a modest space before a new entry. Sometimes it is necessary to leave space for work that will be completed on a separate piece of paper and glued or taped in later. Students might also leave space after a response, so that they can add to it at a later time.

Index. Scientific academic language is important. FOSS strives to have students use precise, accurate vocabulary at all times in their writing and conversations. To help them learn scientific vocabulary, students should set up an index at the end of their notebooks. It is not usually possible for students to enter the words in alphabetical order, as they will be acquired as the course advances. Instead, students could use several pages at the end of the notebook blocked out in 24 squares, and assign one or more letters to each square. Students write the new vocabulary word or phrase in the appropriate square and tag it with the page number of the notebook on which the word is defined. By developing vocabulary in context, students construct meaning through the inquiry process, and by organizing the words in an index, they strengthen their science notebooks as a documentary tool of their science learning. As another alternative, students can also define the word within these squares with the page references.

Table of contents

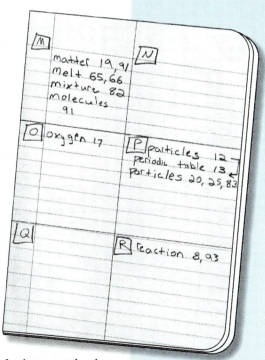

A science notebook index

Science Notebooks in Middle School

Science Notebooks in Middle School

NOTEBOOK COMPONENTS

Four general types of notebook entries, or components, give the science notebook conceptual shape and direction. These structures don't prescribe a step-by-step procedure for how to prepare the notebook, but they do provide some overall guidance. The general arc of an investigation starts with a question or challenge, proceeds with an activity and data acquisition, continues to sense making, and ends with next steps such as reflection and self-assessment.

All four components are not necessary during each class session, but over the course of an investigation, each component will be visited at least once. It may be useful to keep these four components in mind as you systematically guide students through their notebook entries. The components are overviewed here and described in greater detail on the following pages.

Focusing the investigation. Each part of each FOSS investigation includes a focus question, which students transcribe into their notebooks. Focus questions are embedded in the teacher step-by-step instructions and explicitly labeled. The focus question establishes the direction and conceptual challenge for that part of the investigation. For instance, when students investigate the origins of sand and sandstone in the **Earth History Course**, they start by writing,

➤ *Which came first, sand or sandstone?*

The question focuses both students and you on the learning goals for the activity. Students may start by formulating a plan, formally or informally, for answering the focus question. The goal of the plan is to obtain a satisfactory answer to the focus question, which will be revisited and answered later in the investigation.

Data acquisition and organization. After students have established a plan, they collect data. Students can acquire data from carefully planned experiments, accurate measurements, systematic observations, free explorations, or accidental discoveries. It doesn't matter what process produces the data; the critically important point is that students obtain data and record it. It may be necessary to reorganize and display the data for efficient analysis, often by organizing a data table. The data display is key to making sense of the science inquiry.

Making sense of data. Once students have collected and displayed their data, they need to analyze it to learn something about the natural world. In this component of the notebook, students write explanatory statements that answer the focus question. You can formalize this component by asking students to use an established protocol such as a sentence starter, or the explanation can be purely a thoughtful effort by each student. Explanations may be incorrect or incomplete at this point, but students can remedy this during the final notebook entry, when they have an opportunity to continue processing what they've learned. Unfortunately, this piece is often forgotten in the classroom during the rush to finish the lesson and move on. But without sense making and reflection (the final phase of science inquiry), students might see the lesson as a fun activity without connecting the experience to the big ideas that are being developed in the course.

Next-step strategies. The final component of an investigation brings students back to their notebooks by engaging in a next-step strategy, such as reflection and self-assessment, that moves their understanding forward. This component is the capstone on a purposeful series of experiences designed to guide students to understand the concept originally presented in the focus question. After making sense of the data, and making new claims about the topic at hand, students should go back to their earlier thinking and note their changing ideas and new findings. This reflective process helps students cement their new ideas.

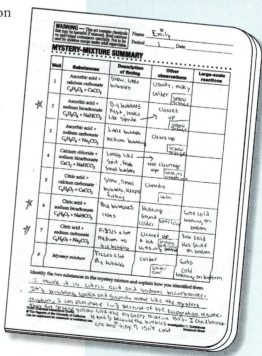

A student organizes and makes sense of data in the Chemical Interactions Course.

Science Notebooks in Middle School

Science Notebooks in Middle School

Focusing the Investigation

Focus question. The first notebook entry in most investigations is the focus question. Focus questions are embedded in the teacher step-by-step instructions and explicitly labeled. You can write the question on the board or project it for students to transcribe into their notebooks. The focus question serves to focus students and you on the inquiry for the day. It is not always answered immediately, but rather hangs in the air while the investigation goes forward. Students always revisit their initial responses later in the investigation.

Quick write. A quick write (or quick draw) can be used in addition to a focus question. Quick writes can be completed on a quarter sheet of paper or an index card so you can collect, review, and return them to students to be taped or glued into their notebooks and used for self-assessment later in the investigation.

In the **Diversity of Life Course**, you ask,

➤ *What is life?*

For a quick write, students write an answer immediately, before instruction occurs. The quick write provides insight into what students think about certain phenomena before you begin instruction. When responding to the question, students should be encouraged to write down their thoughts, even if they don't feel confident in knowing the answer.

Knowing students' preconceptions will help you know what concepts need the most attention during the investigation. Make sure students date their entries for later reference. Read through students' writing and tally the important points to focus on. Quick writes should not be graded.

Planning. After students enter the focus question or complete a quick write in their notebooks, they plan their investigation. (In some investigations, planning is irrelevant to the task at hand.) Planning may be detailed or intuitive, formal or informal, depending on the requirement of the investigation. Plans might include lists (including materials, things to remember), step-by-step procedures, and experimental design. Some FOSS notebook masters guide students through a planning process specific to the task at hand.

Lists. Science notebooks often include lists of things to think about, materials to get, or words to remember. A materials list is a good organizer that helps students anticipate actions they will take. A list of variables to be controlled clarifies the purpose of an experiment. Simple lists of dates for observations or of the people responsible for completing a task may be useful.

Step-by-step procedures. Middle school students need to develop skills for writing sequential procedures. For example, in the **Chemical Interactions Course**, students write a procedure to answer these questions.

➤ *Is there a limit to the amount of substance that will dissolve in a certain amount of liquid?*

➤ *If so, is the amount that will dissolve the same for all substances?*

Students need to recall what they know about the materials, develop a procedure for accurately measuring the amount of a substance that is added to the water, and agree on a definition of "saturated." To check a procedure for errors or omissions, students can trade notebooks and attempt to follow another student's instructions to complete the task.

Experimental design. Some work with materials requires a structured experimental plan. In the **Planetary Science Course**, students pursue this focus question.

➤ *Are Moon craters the result of volcanoes or impacts?*

Students plan an experiment to determine what affects the size and shape of craters on the Moon. They use information they gathered during the open exploration of craters made in flour to develop a strategy for evaluating the effect of changing the variable of a projectile's height or mass. Each lab group agrees on which variable they will change and then designs a sound experimental procedure that they can refer to during the active investigation.

Science Notebooks in Middle School

Science Notebooks in Middle School

Data Acquisition and Organization

Because observation is the starting point for answering the focus question, data records should be

- clearly related to the focus question;
- accurate and precise;
- organized for efficient reference.

Data handling can have two subcomponents: data acquisition and data display. Data acquisition is making and recording observations (measurements). The data record can be composed of words, phrases, numbers, and drawings. Data display reorganizes the data in a logical way to facilitate thinking. The display can be a graph, chart, calendar, or other graphic organizer.

Early in a student's experience with notebooks, the record may be disorganized and incomplete, and the display may need guidance. The FOSS notebook masters are designed to help students with data collection and organization. You may initially introduce conventional data-display methods, such as those found in the FOSS notebook masters, but soon students will need opportunities to independently select appropriate data displays. As students become more familiar with collecting and organizing data, you might have them create their own records. With practice, students will become skilled at determining what form of recording to use in various situations, and how best to display the data for analysis.

Narratives. For most students, the most intuitive approach to recording data is narrative—using words, sentence fragments, and numbers in a more or less sequential manner. As students make a new observation, they record it below the previous entry, followed by the next observation, and so on. Some observations, such as a record of weather changes in the **Weather and Water Course** or the interactions of organisms in miniecosystems in the **Populations and Ecosystems Course**, are appropriately recorded in narrative form.

Drawings. A picture is worth a thousand words, and a labeled picture is even more useful. When students use a microscope to discover cells in the *Elodea* leaf and observe and draw structures of microorganisms in the **Diversity of Life Course**, a labeled illustration is the most efficient way to record data.

Charts and tables. An efficient way to record many kinds of data is a chart or table. How do you introduce this skill into the shared knowledge of the classroom? One way is to call for attention during an investigation and demonstrate how to perform the operation. Or you can let students record the data as they like, and observe their methods. There may be one or more groups that invent an appropriate table. During processing time, ask this group to share its method with the class. If no group has spontaneously produced an effective table, you might challenge the class to come up with an easier way to display the data, and turn the skill-development introduction into a problem-solving session.

With experience, students will recognize when a table or chart is appropriate for recording data. When students make similar observations on a series of objects, such as rock samples in the **Earth History Course**, a table with columns is an efficient way to organize observations for easy comparison.

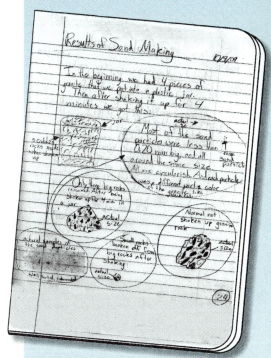

Drawing and artifact from the Earth History Course

Artifacts. Occasionally an investigation will produce two-dimensional artifacts that students can tape or glue directly into a science notebook. The mounted flower parts in the **Diversity of Life Course** and the sand samples card from the **Earth History Course** can become a permanent part of the record of learning.

Graphs and graphic tools. Reorganizing data into logical, easy-to-use graphic tools is typically necessary for data analysis. Graphs allow easy comparison (bar graph), quick statistical analysis of frequency data (histogram or line plot), and visual confirmation of a relationship between variables (two-coordinate graph). The **Variables and Design Course** offers many opportunities for students to collect data and organize the data into graphs. Students collect data from air trolleys traveling at different speeds, graph the data, and use the resulting graphs to understand how slope of a motion graph can indicate speed. Other graphic tools, such as Venn diagrams, pie charts, and concept maps, help students make connections.

Science Notebooks in Middle School

Making Sense of Data

After collecting and organizing data, the student's next task is to answer the focus question. Students can generate an explanation as an unassisted narrative, but in many instances you might need to use supports such as the FOSS notebook masters to guide the development of a coherent and complete response to the question. Several other support structures for sense making are described below.

Development of vocabulary. Vocabulary is better introduced after students have experienced the new word(s) in context. This sequence provides a cognitive basis for students to connect accurate and precise language to their real-life experiences. Lists of new vocabulary words in the index reinforce new words and organize them for easy reference.

Data analysis. Interpreting data requires the ability to look for patterns, trends, outliers, and potential causes. Students should be encouraged to develop a habit of looking for patterns and relationships within the data collected. Frequently, this is accomplished by creating a graph with numerical data. In the **Populations and Ecosystems Course**, students review field data acquired by ecologists at Mono Lake to determine how biotic and abiotic factors affect the populations of organisms found in the lake.

Graphic organizers. Students can benefit from organizers that help them look at similarities and differences. A compare-and-contrast chart can help students make a transition from their collected data and experiences to making and writing comparisons. It is sometimes easier for students to use than a Venn diagram, and is commonly referred to as a box-and-T chart (as popularized in *Writing in Science: How to Scaffold Instruction to Support Learning*, listed in the Bibliography section).

In this strategy, students draw a box at the top of the notebook page and label it "similar" or "same." On the bottom of the notebook page, they draw a *T*. At the top of each wing of the *T*, they label the objects being compared. Students look at their data, use the *T* to identify differences for each item, and use the "similar" box to list all the characteristics that the two objects have in common. For example, a box-and-T chart comparing characteristics of extrusive and intrusive igneous rocks in the **Earth History Course** might look like this.

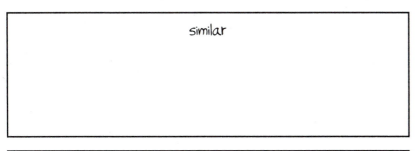

Students can use the completed box-and-T chart to begin writing comparisons. It is usually easier for students to complete their chart on a separate piece of paper, so they can fill it in as they refer to their data. They affix the completed chart into their notebooks after they have made their comparisons.

Claims and evidence. A claim is an assertion about how the natural world works. Claims should always be supported by evidence—statements that are directly correlated with data. The evidence should refer to specific observations, relationships that are displayed in graphs, tables of data that show trends or patterns, dates, measurements, and so on. A claims-and-evidence construction is a sophisticated, rich display of student learning and thinking. It also shows how the data students collected is directly connected to what they learned.

Science Notebooks in Middle School

Science Notebooks in Middle School

Frames and prompts. One way to get students to organize their thinking is by providing sentence frames for them to complete.

- I used to think _____, but now I think _____.
- The most important thing to remember about Moon phases is _____.
- One new thing I learned about adaptation is _____.

Prompts also direct students to the content they should be thinking about, but provide more latitude for generating responses. For students who are learning English or who struggle with writing, assistive structures like sentence frames can help them communicate their thinking while they learn the nuances of science writing. The prompts used most often in the FOSS notebook masters take the form of questions for students to answer. In the **Weather and Water Course**, students answer the quick-write question

➤ What causes seasons?

After modeling an Earth/Sun system and reviewing solar angle and solar concentration, students revisit their quick write to revise and expand on their original explanations.

- I used to think seasons were caused by _____, but now I know _____.

Careful prompts scaffold students by helping them communicate their thinking but do not do the thinking for them. As students progress in communication ability, you might provide frames less frequently.

Conclusions and predictions. At the end of an investigation (major conceptual sequence), it might be appropriate for students to write a summary to succinctly communicate what they have learned. This is where students can make predictions based on their understanding of a principle or relationship. For instance, after completing the investigation of condensation and dew point in the **Weather and Water Course**, a student might predict the altitude at which clouds would form, based on weather-balloon data. Or, after examining ecosystem interactions between biotic and abiotic factors in the **Populations and Ecosystems Course**, students will predict how various human interactions could affect the ecosystem. The conclusion or prediction will frequently indicate the degree to which a student can apply new knowledge to real-world situations. A prediction can also be the springboard for further inquiry.

Generating new questions. Does the investigation connect to a student's personal interests? Does the outcome suggest a question or pique a student's curiosity? The science classroom is most exciting when students are generating their own questions for further investigation based on class or personal experiences. The notebook is an excellent place to capture students' musings and record thoughts that might otherwise be lost.

Science Notebooks in Middle School

Next-Step Strategies

The goal of the FOSS curriculum is for students to develop accurate, durable knowledge of the science content under investigation. Students' initial conceptions are frequently incomplete or confused, requiring additional thought to become fully functional. The science notebook is a useful place to guide reflection and revision. Typically students commit their understanding in writing and reflect in three locations.

- Explanatory narratives in notebooks
- Response sheets incorporated into the notebook
- Written work on I-Checks

These three categories of written work provide information about student learning for you *and* a record of thinking for students that they can reflect on and revise. Scientists constantly refine and clarify their ideas about how the natural world works. They read scientific articles, consult with other scientists, and attend conferences. They incorporate new information into their thinking about the subject they are researching. This reflective process can result in deeper understanding or a complete revision of thinking.

After completing one of the expositions of knowledge—a written conclusion, response sheet, or benchmark assessment—students should receive additional instruction or information via a next-step strategy. They will use this information later to complete self-assessment by reviewing their original written work, making judgments about its accuracy and completeness, and writing a revised explanation. You can use any of a number of techniques for providing the additional information to students.

- Group compare-and-share discussion
- Think/pair/share reading
- Whole-class critique of an explanation by an anonymous student
- Identifying key points for a class list
- Whole-class discussion of a presentation by one student

After one of the information-generating processes, students compare the "best answer" to their own answer and rework their explanations if they can no longer defend their original thinking. The revised statement of the science content can take one of several forms.

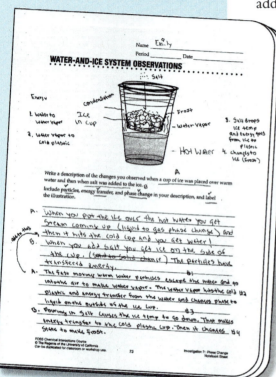

A student's revised work for the Chemical Interactions Course

Students might literally revise the original writing, crossing out extraneous or incorrect bits, inserting new or improved information, and completing the passage. At other times, students might reflect on their original work and, after drawing and dating a line of learning (see below), might redraft their explanation from scratch, producing their best explanation of the concept.

During these self-assessment processes, students have to think actively about every aspect of their understanding of the concept and organize their thoughts into a coherent, logical narrative. The learning that takes place during this process is powerful. The relationships between the several elements of the concept become unified and clarified.

The notebook is the best tool for students when preparing for benchmark assessment, such as an I-Check or posttest. Students don't necessarily have the study skills needed to prepare on their own, but using teacher-guided tasks such as key points and traffic lights will turn the preparation process into a valuable exercise. These same strategies can be used after a benchmark assessment when you identify further areas of confusion or misconceptions you want to address with students. Here are four helpful next-step, or self-assessment, strategies.

Line of learning. One technique many teachers find useful in the reflective process is the line of learning. After students have conducted an investigation and entered their initial explanations, they draw and date a line under their original work. As students share ideas and refine their thinking during class discussion, additional experimentation, reading, and teacher feedback, encourage them to make new entries under the line of learning, adding to or revising their original thinking. If the concept is elusive or complex, a second line of learning, followed by more processing and revising, may be appropriate.

The line of learning is a reminder to students that learning is an ongoing process with imperfect products. It points out places in that process where a student made a stride toward full understanding. And the psychological security provided by the line of learning reminds students that they can always draw another line of learning and revise their thinking again. The ability to look back in the science notebook and see concrete evidence of learning gives students confidence and helps them become critical observers of their own learning.

A line of learning used with the Planetary Science Course

Science Notebooks in Middle School

Science Notebooks in Middle School

Traffic lights. In the traffic-lights strategy, students use color to self-assess and indicate how well they understand a concept that they are learning. Green means that the student feels that he or she has a good understanding of the concept. Yellow means that the student is still a bit unsure about his or her understanding. Red means that the student needs help; he or she has little or no understanding of the concept. Students can use colored pencils, markers, colored dots, or colored cards to indicate their understanding. They can mark their own work and then indicate their level of understanding by a show of hands or by holding up colored cards. This strategy gives students practice in self-assessment and helps you monitor students' current understanding. You should follow up by looking at student work to ensure that they actually do understand the content that they marked with green.

Three C's. Another approach to revision is to apply the three C's—confirm, correct, complete—to the original work. Students indicate ideas that were correct with a number or a color, code statements needing correction with a second number or color, and assign a third number or color to give additional information that completes the entry.

Key points. Students do not necessarily connect the investigative experience with the key concepts and processes taught in the lesson. It is essential to give students an opportunity to reflect on their experiences and find meaning in those experiences. They should be challenged to use their experiences and data to either confirm or reject their current understanding of the natural world. As students form supportable ideas about a concept, those ideas should be noted as key points, posted in the room, and written in their notebooks. New evidence, to support or clarify an idea, can be added to the chart as the course progresses. If an idea doesn't hold up under further investigation, a line can be drawn through the key point to indicate a change in thinking. A key-points activity is embedded near the end of each investigation to help students organize their thinking and prepare for benchmark assessment.

USING NOTEBOOKS TO IMPROVE STUDENT LEARNING

Notebook entries should not be graded. Research has shown that more learning occurs when students get only comments on written work in their notebooks, not grades or a combination of comments and grades.

If your school district requires a certain number of grades each week, select certain work products that you want to grade and have students turn in that work separate from the notebook. After grading, return the piece to students to insert into their notebooks, so that all their work is in one place.

It may be difficult to stop using grades or a rubric for notebook assessment. But providing feedback that moves learning forward, however difficult, has benefits that make it worth the effort. The key to using written feedback for formative assessment is to make feedback timely and specific, and to provide time for students to act on the feedback by revising or correcting work right in their own notebook.

Teacher Feedback

Student written work often exposes weaknesses in understanding—or so it appears. It is important for you to find out if the flaw results from poor understanding of the science or from imprecise communication. You can use the notebook to provide two types of feedback to the student: to ask for clarification or additional information, and to ask probing questions that will help students move forward in their thinking. Respecting the student's space is important, so rather than writing directly in the notebook, attach a self-stick note, which can be removed after the student has taken appropriate action.

The most effective forms of feedback relate to the content of the work. Here are some examples.

> ➤ *You wrote that seasons are caused by Earth's tilt. Does Earth's tilt change during its orbit?*
>
> ➤ *What evidence can you use to support your claim that Moon craters are caused by impacts? Hint: Think of our experiments in class.*

Nonspecific feedback, such as stars, pluses, smiley faces, and "good job!", or ambiguous critiques, such as "try again," "put more thought into this," and "not enough," are less effective and should not be used. Feedback that guides students to think about the content of their work and gives suggestions for how to improve are productive instructional strategies.

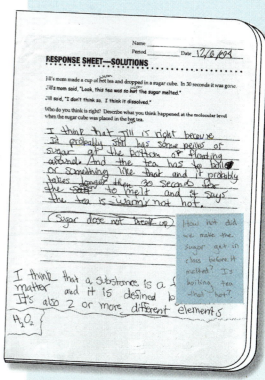

Feedback given during the Chemical Interactions Course

Science Notebooks in Middle School

Here are some appropriate generic feedback questions to write or use verbally while you circulate in the class.

➤ *What vocabulary have you learned that will help you describe _____ ?*

➤ *Can you include an example from class to support your ideas?*

➤ *Include more detail about _____ .*

➤ *Check your data to make sure this is accurate.*

➤ *What do you mean by _____ ?*

➤ *When you record your data, what unit should you use?*

When students return to their notebooks and respond to the feedback, you will have additional information to help you discriminate between learning and communication difficulties. Another critical component of teacher feedback is providing comments to students in a timely manner, so that they can review their work before engaging in benchmark assessment or moving on to other big ideas in the course.

In middle school, you face the challenge of having a large number of students. This may mean collecting a portion of students' notebooks on alternate days. Set a specific focus for your feedback, such as a data table or conclusion, so you aren't trying to look at everything every time.

To help students improve their writing, you might have individuals share notebook entries aloud in their collaborative groups, followed by feedback from a partner or the group. This valuable tool must be very structured to create a safe environment, including ground rules about acceptable feedback and comments.

A good way to develop these skills is to model constructive feedback with the class, using a student-work sample from a notebook. Use a sample from a previous year with the name and any identifying characteristics removed. Project it for the class to practice giving feedback.

Formative Assessment

With students recording more of their thinking in an organized notebook, you have a tool to better understand the progress of students and any misconceptions that are typically not revealed until the benchmark assessment. One way to monitor student progress is during class while they are responding to a prompt. Circulate from group to group, and read notebook entries over students' shoulders. This is a good time to have short conversations with individuals or small groups to gain information about the level of student understanding. Take care to respect the privacy of students who are not comfortable sharing their work during the writing process.

If you want to look at work that is already completed in the notebook, ask students to open their notebooks to the page that you want to review and put them in a designated location. Or consider having students complete the work on a separate piece of paper or an index card. Students can leave a blank page in their notebooks, or label it with a header as a placeholder, until they get the work back and tape it or glue it in place. This makes looking at student work much easier, and the record of learning that the student is creating in the notebook remains intact.

When time is limited, you might select a sample of students from each class, alternating the sample group each time, to get a representative sample of student thinking. This is particularly useful following a quick write.

Once you have some information about student thinking, you can make teaching decisions about moving ahead to a benchmark assessment, going back to a previous concept, or spending more time making sense of a concept. Benchmark assessments can also be used as formative assessment. You might choose to administer an I-Check, score the assessment to find problem areas, and then revisit critical concepts before moving on to the next investigation. Students can use reflection and self-assessment techniques to revisit and build on their original exam responses.

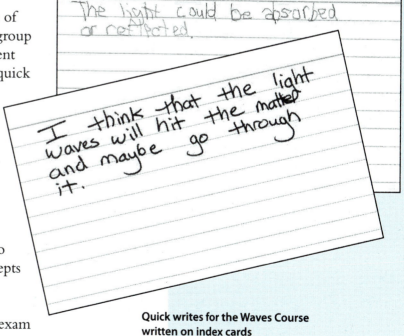

Quick writes for the Waves Course written on index cards

Science Notebooks in Middle School

Science Notebooks in Middle School

DERIVATIVE PRODUCTS

On occasion, you might ask students to produce science projects of various kinds: summary reports, detailed explanations, end-of-course projects, oral reports, or posters. Students should use their notebooks as a reference when developing their reports. You could ask them to make a checklist of science concepts and pieces of evidence, with specific page references, extracted from their notebooks. They can then use this checklist to ensure that all important points have been included in the derivative work.

The process of developing a project has feedback benefits, too. While students are developing projects using their notebooks, they have the opportunity to self-monitor the organization and content of the notebook. This offers valuable feedback on locating and extracting useful information. You might want to discuss possible changes students would make next time they start a new science notebook.

Homework is another form of derivative product, as it is an extension of the experimentation started in class. Carefully selected homework assignments enhance students' science learning. Homework suggestions and/or extension activities are included at the end of each investigation. For example, in the **Heredity and Adaptation Course**, after using an online activity in class to predict genetic variation, students are asked to complete a follow-up online simulation as homework. In the **Electromagnetic Force Course**, after students test properties of magnets in class, they are asked to look for examples of magnets in household objects outside of the classroom.

Science-Centered Language Development in Middle School

Science-Centered Language Development in Middle School

Reading and writing are inextricably linked to the very nature and fabric of science, and, by extension, to learning science.

Stephen P. Norris and Linda M. Phillips, "How Literacy in Its Fundamental Sense Is Central to Scientific Literacy"

Contents

Introduction	C1
The Role of Language in Scientific and Engineering Practices	C3
Speaking and Listening Domain	C6
Writing Domain	C12
Reading Domain	C18
Science-Vocabulary Development	C26
English-Language Development	C31
References	C43

INTRODUCTION

In this chapter, we explore the ways reading, writing, speaking, and listening are interwoven in effective science instruction at the secondary level. To engage fully in the enterprise of science and engineering, students must record and communicate observations and explanations, and read about and discuss the discoveries and ideas of others. This becomes increasingly challenging at the secondary level. Texts become more complex; writing requires fluency of academic language, including domain-specific vocabulary. Here we identify strategies that support sense making. The active investigations, student science notebooks, *FOSS Science Resources* readings, multimedia, and formative assessments provide rich contexts in which students develop and exercise thinking processes and communication skills. Students develop scientific literacy through experiences with the natural world around them in real and authentic ways and use language to inquire, process information, and communicate their thinking about the objects, organisms, and phenomena they are studying. We refer to the acquisition and building of language skills necessary for scientific literacy as science-centered language development.

Full Option Science System   *Copyright © The Regents of the University of California*

Science-Centered Language Development in Middle School

Language plays two crucial roles in science learning: (1) it facilitates the communication of conceptual and procedural knowledge, questions, and propositions (external; public), and (2) it mediates thinking, a process necessary for understanding (internal; private). These are also the ways scientists use language: to communicate with one another about their inquiries, procedures, and understandings; to transform their observations into ideas; and to create meaning and new ideas from their work and the work of others. For students, language development is intimately involved in their learning about the natural world. Active-learning science provides a real and engaging context for developing literacy; language-arts skills and strategies support conceptual development and scientific and engineering practices. For example, the skills and strategies used for reading comprehension, writing expository text, and oral discourse are applied when students are recording their observations, making sense of science content, and communicating their ideas. Students' use of language improves when they discuss, write, and read about the concepts explored in each investigation.

We begin our exploration of science and language by focusing on language functions and how specific language functions are used in science to facilitate information acquisition and processing (thinking). Then we address issues related to the specific language domains—speaking and listening, writing, and reading. Each section addresses

- how skills in that domain are developed and exercised in FOSS science investigations;
- literacy strategies that are integrated purposefully into the FOSS investigations; and
- suggestions for additional literacy strategies that both enhance student learning in science and develop or exercise academic-language skills.

Following the domain discussions is a section on science-vocabulary development, with scaffolding strategies for supporting all learners. The last section covers language-development strategies specifically for English learners.

> **NOTE**
> The term *English learners* refers to students who are learning to understand English. This includes students who speak English as a second language and native English speakers who need additional support to use language effectively.

THE ROLE OF LANGUAGE IN SCIENTIFIC AND ENGINEERING PRACTICES

Language functions are the purpose for which speech or writing is used and involve both vocabulary and grammatical structure. Understanding and using language functions appropriately is important in effective communication. Students use numerous language functions in all disciplines to mediate communication and facilitate thinking (e.g., they plan, compare, discuss, apply, design, draw, and provide evidence).

In science, language functions facilitate scientific and engineering practices. For example, when students are *collecting data*, they are using language functions to identify, label, enumerate, compare, estimate, and measure. When students are *constructing explanations*, they are using language functions to analyze, communicate, discuss, evaluate, and justify.

A Framework for K–12 Science Education (National Research Council 2012) states that "Students cannot comprehend scientific practices, nor fully appreciate the nature of scientific knowledge itself, without directly experiencing the practices for themselves." Each of these scientific and engineering practices uses multiple language functions. Often, these language functions are part of an internal dialogue weighing the merits of various explanations—what we call thinking. The more language functions with which we are facile, the more effective and creative our thinking can be.

The scientific and engineering practices are listed below, along with a sample of the language functions that are exercised when students are effectively engaged in that practice. (Practices are bold; language functions are italic.)

Asking questions and defining problems

- *Ask* questions about objects, organisms, systems, and events in the natural and human-made world (science).
- *Ask* questions to *define* and *clarify* a problem, *determine criteria* for solutions, and *identify* constraints (engineering).

Planning and carrying out investigations

- *Plan* and conduct investigations in the laboratory and in the field to gather appropriate data (*describe* procedures, *determine* observations to *record*, *decide* which variables to control) or to gather data essential for *specifying* and *testing* engineering designs.

Examples of Language Functions
Analyze
Apply
Ask
Clarify
Classify
Communicate
Compare
Conclude
Construct
Critique
Describe
Design
Develop
Discuss
Distinguish
Draw
Enumerate
Estimate
Evaluate
Experiment
Explain
Formulate
Generalize
Group
Identify
Infer
Interpret
Justify
Label
List
Make a claim
Measure
Model
Observe
Organize
Plan
Predict
Provide evidence
Reason
Record
Represent
Revise
Sequence
Solve
Sort
Strategize
Summarize
Synthesize

Science-Centered Language Development in Middle School

Analyzing and interpreting data

- Use a range of tools (numbers, words, tables, graphs, images, diagrams, equations) to *organize* observations (data) in order to *identify* significant features and patterns.

Developing and using models

- Use models to help *develop explanations, make predictions*, and *analyze* existing systems, and *recognize* strengths and limitations of the models.

Using mathematics and computational thinking

- Use mathematics and computation to *represent* physical variables and their relationships.

Constructing explanations and designing solutions

- *Construct* logical explanations of phenomena, or *propose solutions* that incorporate current understanding or a model that represents it and is consistent with the available evidence.

Engaging in argument from evidence

- *Defend* explanations, *formulate evidence* based on data, *examine* one's own understanding in light of evidence offered by others, and collaborate with peers in searching for explanations.

Obtaining, evaluating, and communicating information

- *Communicate* ideas and the results of inquiry—orally and in writing—with tables, diagrams, graphs, and equations and in *discussion* with peers.

Research supports the claim that when students are intentionally using language functions in thinking about and communicating in science, they improve not only science content knowledge, but also language-arts and mathematics skills (Ostlund, 1998; Lieberman and Hoody, 1998). Language functions play a central role in science as a key cognitive tool for developing higher-order thinking and problem-solving abilities that, in turn, support academic literacy in all subject areas.

Here is an example of how an experienced teacher can provide an opportunity for students to exercise language functions in FOSS. In the **Planetary Science Course**, one piece of content we expect students to have acquired by the end of the course is

- The lower the angle at which light strikes a surface, the lower the density of the light energy.

The scientific practices the teacher wants the class to focus on are *developing and using models* and *constructing explanations*.

The language functions students will exercise while engaging in these scientific practices include *comparing, modeling, analyzing*, and *explaining*. The teacher understands that these language functions are appropriate to the purpose of the science investigation and support the *Common Core State Standards for English Language Arts and Literacy in Science* (CCSS), in which grades 6–8 students will "write arguments focused on discipline-specific content . . . support claim[s] with logical reasoning and relevant, accurate data and evidence that demonstrate an understanding of the topic" (National Governors Association Center for Best Practices, Council of Chief State School Officers, 2010).

▶ **CCSS NOTE**
This example supports
CCSS.ELA-Literacy.WHST.6–8.1.b.

- Students will *compare* the area covered by the same beam of light (from a flashlight) at different angles to *explain* the relationship between the angle and density of light energy.

The teacher can support the use of language functions by providing structures such as sentence frames.

- As _____, then _____.

  As the angle increases, then the light beam becomes smaller and more circular.

- When I changed _____, then _____.

  When I changed the angle of the light beam, then the concentration of light hitting the floor changed.

- The greater/smaller _____, the _____.

  The smaller the angle of the light beam, the more the light beam spread out.

- I think _____, because _____.

  I think the smaller spot of light receives more energy than the larger spot because the light concentration is greatest when light shines directly down on a surface and there is no beam spreading.

Science-Centered Language Development in Middle School

Science-Centered Language Development in Middle School

SPEAKING AND LISTENING DOMAIN

The FOSS investigations are designed to engage students in productive oral discourse. Talking requires students to process and organize what they are learning. Listening to and evaluating peers' ideas calls on students to apply their knowledge and to sharpen their reasoning skills. Guiding students in instructive discussions is critical to the development of conceptual understanding of the science content and the ability to think and reason scientifically. It also addresses a key middle school CCSS Speaking and Listening anchor standard that students "engage effectively in a range of collaborative discussions (one-on-one, in groups, and teacher-led) with diverse partners on [grade-level] topics, texts, and issues, building on others' ideas and expressing their own clearly."

▶ **CCSS NOTE**
This example supports CCSS.ELA-Literacy.SL.6.1, CCSS.ELA-Literacy.SL.7.1, and CCSS.ELA-Literacy.SL.8.1.

FOSS investigations start with a discussion—a review to activate prior knowledge, presentation of a focus question, or a challenge to motivate and engage active thinking. During the active investigation, students talk with one another in small groups, share their observations and discoveries, point out connections, ask questions, and start to build explanations. The discussion icon in the sidebar of the *Investigations Guide* indicates when small-group discussions should take place.

Throughout the activity, the *Investigations Guide* indicates where it is appropriate to pause for whole-class discussions to guide conceptual understanding. The *Investigations Guide* provides you with discussion questions to help stimulate student thinking and support sense making. At times, it may be beneficial to use sentence frames or standard prompts to scaffold the use of effective language functions and structures. Allowing students a few minutes to write in their notebooks prior to sharing their answers also helps those who need more time to process and organize their thoughts.

▶ **NOTE**
Additional notebook strategies can be found in the Science Notebooks in Middle School chapter in *Teacher Resources* and online at www.FOSSweb.com.

On the following pages are some suggestions for providing structure to those discussions and for scaffolding productive discourse when needed. Using the protocols that follow will ensure inclusion of all students in discussions.

Partner and Small-Group Discussion Protocols

Whenever possible, give students time to talk with a partner or in a small group before conducting a whole-class discussion. This provides all students with a chance to formulate their thinking, express their ideas, practice using the appropriate science vocabulary, and receive input from peers. Listening to others communicate different ways of thinking about the same information from a variety of perspectives helps students negotiate the difficult path of sense making for themselves.

Dyads. Students pair up and take turns either answering a question or expressing an idea. Each student has 1 minute to talk while the other student listens. While student A is talking, student B practices attentive listening. Student B makes eye contact with student A, but cannot respond verbally. After 1 minute, the roles reverse.

Here's an example from the **Chemical Interactions Course**. After reviewing the results of eight reactions recorded in their notebooks, you ask students to pair up and take turns sharing which two substances they think constitute the mystery mixture and their reasons for selecting those two. The language objective is for students to compare their test results and make inferences based on their observations and what they know about chemical reactions (orally and in writing). These sentence frames can be written on the board to scaffold student thinking and conversation.

- I think the two substances in the mystery mixture are _____ and _____.

- My evidence is _____.

Partner parade. Students form two lines facing each other. Present a question, an idea, an object, or an image as a prompt for students to discuss. Give them 1 minute to greet the person in front of them and discuss the prompt. After 1 minute, call time. Have the first student in one of the lines move to the end of the line, and have the rest of the students in that line shift one step sideways so that everyone has a new partner. (Students in the other line do not move.) Give students a new prompt to discuss for 1 minute with their new partners. This can also be done by having students form two concentric circles. After each prompt, the inner circle rotates.

For example, when students are just beginning the **Earth History Course** investigation on igneous rock, you may want to assess prior knowledge about Earth's layers. Give each student a picture from an assortment of related images such as volcanoes, magma, a diagram of Earth's layers, crystals, and so forth, and have students line up facing

---

Partner and Small-Group Discussion Protocols
- *Dyads*
- *Partner parade*
- *Put in your two cents*

Science-Centered Language Development in Middle School

each other in two lines or in concentric circles. For the first round, ask, "What do you observe in the image on your card?" For the second round, ask, "What can you infer from the image?" For the third round, ask, "What questions do you have about the image?" The language objective is for students to describe their observations, infer how the landform formed, and reflect upon and relate any experiences they may have had with a similar landform. These sentence frames can be used to scaffold student discussion.

- I observe _____, _____, and _____.
- I think this shows _____ because _____.
- I wonder _____.

Put in your two cents. For small-group discussions, give each student two pennies or similar objects to use as talking tokens. Each student takes a turn putting a penny in the center of the table and sharing his or her idea. Once all have shared, each student takes a turn putting in the other penny and responding to what others in the group have said. For example,

- I agree (or don't agree) with _____ because _____.

Here's an example from the **Diversity of Life Course**. In their notebooks, students have recorded the amount of water lost from their vials containing celery with and without leaves. They discover a discrepancy in the amount of water lost and the mass of the celery. Where did the water go? Students are struggling to form an explanation. The language objective is for students to compare their results and infer that there is a relationship between the amount of water lost and the number of leaves the celery has. You give each student two pennies, and in groups of four, they take turns putting in their two cents. For the first round, each student answers the question "Where did the water go?" They use the frame

- I think the water _____.
- My evidence is _____.

On the second round, each student states whether he or she agrees or disagrees with someone else in the group and why, using the sentence frame.

Whole-Class Discussion Supports

The whole-class discussion is a critical part of sense making. After students have had the active learning experience and have talked with their peers in pairs and/or small groups, sharing their observations with the whole class sets the stage for developing conventional explanatory models. Discrepant events, differing results, and other surprises are discussed, analyzed, and resolved. It is important that students realize that science is a process of finding out about the world around them. This is done through asking questions, testing ideas, forming explanations, and subjecting those explanations to logical scrutiny, that is, argumentation. Leading students through productive discussion helps them connect their observations and the abstract symbols (words) that represent and explain those observations. Whole-class discussion also provides an opportunity for you to interject an accurate and precise verbal summary as a model of the kind of thinking you are seeking. Facilitating effective whole-class discussions takes skill, practice, a shared set of norms, and patience. In the long run, students will have a better grasp of the content and will improve their ability to think independently and communicate effectively.

Norms should be established so that students know what is expected during science discussions.

- Science content and practices are the focus.
- Everyone participates (speaking and listening).
- Ideas and experiences are shared, accepted, and valued. Everyone is respectful of one another.
- Claims are supported by evidence.
- Challenges (debate and argument) are part of the quest for complete understanding.

A variety of whole-class discussion techniques can be introduced and practiced during science instruction that address the CCSS Speaking and Listening standards for students to "present claims and findings [e.g., argument, narrative, informative, summary presentations], emphasizing salient points in a focused, coherent manner with relevant evidence, sound valid reasoning, and well-chosen details; use appropriate eye contact, adequate volume, and clear pronunciation."

For example, during science talk, students are reminded to practice attentive listening, stay focused on the speaker, ask questions, and respond appropriately. In addition, in order for students to develop and practice their reasoning skills, they need to know the language forms

Whole-Class Discussion Supports
- *Sentence frames*
- *Guiding questions*

TEACHING NOTE

Let students know that scientists change their minds based on new evidence. It is expected that students will revise their thinking, based on evidence presented in discussions.

▶ **CCSS NOTE**
This example supports CCSS.ELA-Literacy.SL.6.4, CCSS.ELA-Literacy.SL.7.4, and CCSS.ELA-Literacy.SL.8.4, (grade 8 quoted here).

Science-Centered Language Development in Middle School

Science-Centered Language Development in Middle School

> **TEACHING NOTE**
>
> *Encourage "science talk." Allow time for students to engage in discussions that build on other students' observations and reasoning. After an investigation, use a teacher- or student-generated question, and either just listen or facilitate the interaction with questions to encourage expression of ideas among students.*

and structures and the behaviors used in evidence-based debate and argument, such as using data to support claims, disagreeing respectfully, and asking probing questions (Winokur and Worth, 2006).

Explicitly model the language structures appropriate for active discussions, and encourage students to use them when responding to guiding questions and during science talks.

Sentence frames. These samples can be posted as a scaffold as students develop their reasoning and oral participation skills.

- I think _____, because _____.
- I predict _____, because _____.
- I claim _____; my evidence is _____.
- I agree with _____ that _____.
- My idea is similar/related to _____'s idea.
- I learned/discovered/heard that _____.
- <Name> explained _____ to me.
- <Name> shared _____ with me.
- We decided/agreed that _____.
- Our group sees it differently, because _____.
- We have different observations/results. Some of us found that _____. One group member thinks that _____.
- We had a different approach/idea/solution/answer: _____.

Guiding questions. The Investigations Guide provides questions to help concentrate student thinking on the concepts introduced in the investigation. Guiding questions should be used during the whole-class discussion to facilitate sense making. Here are some other open-ended questions that help guide student thinking and promote discussion.

- What did you notice when _____?
- What do you think will happen if _____?
- How might you explain _____? What is your evidence?
- What connections can you make between _____ and _____?

Whole-Class Discussion Protocols

The following tried-and-true participation protocols can be used to enhance whole-class discussions. The purpose of these protocols is to increase meaningful participation by giving all students access to the discussion, allowing students time to think (process), and providing a context for motivation and engagement.

Think-pair-share. When asking for a response to a question posed to the class, give students a minute to think silently. Then, have students pair up with a partner to exchange thoughts before you call on a student to share his or her ideas with the whole class.

Pick a stick. Write each student's name on a craft stick, and keep the sticks handy in a cup at the front of the room. When asking for responses, randomly pick a stick, and call on that student to start the discussion. Continue to select sticks as you continue the discussion. Your name can also be on a stick in the cup. To keep students on their toes, put the selected sticks into a smaller cup hidden inside the larger cup out of view of students. That way students think they may be called again.

Whip around. Each student takes a quick turn sharing a thought or reaction. Questions are phrased to elicit quick responses that can be expressed in one to five words (e.g., "Give an example of a stored-energy source." "What does the word *heat* make you think of?").

Commit and toss. Have students write a response to a question or prompt on a loose piece of paper (Keeley, 2008). Next, tell everyone to crumple up the paper into a ball and toss it to another student. Continue tossing for a few minutes, and then call for students to stop, grab a ball, and read the response silently. Responses can then be shared with partners, small groups, or the whole class. This activity allows students to answer anonymously, so they may be willing to share their thinking more openly.

Group posters. Have small groups design and graphically record their investigation data and conclusions on a quickly generated poster to share with the whole class.

Whole-Class Discussion Protocols
- *Think-pair-share*
- *Pick a stick*
- *Whip around*
- *Commit and toss*
- *Group posters*

Cup within a cup pick-a-stick container

Science-Centered Language Development in Middle School

Science-Centered Language Development in Middle School

> **NOTE**
> For more information about supporting science-notebook development, see the Science Notebooks in Middle School chapter.

> **CCSS NOTE**
> This example supports CCSS.ELA-Literacy.W.10.

WRITING DOMAIN

Information processing is enhanced when students engage in informal writing. When allowed to write expressively without fear of being scorned for incorrect spelling or grammar, students are more apt to organize and express their thoughts in different ways that support their own sense making. Writing in science promotes use of science and engineering practices, thereby developing a deeper engagement with the science content. This type of informal writing also provides a springboard for more formal derivative science writing (Keys, 1999).

Science Notebooks

The science notebook is an effective tool for enhancing learning in science and exercising various forms of writing. Notebooks provide opportunities both for expressive writing (students craft explanatory narratives that make sense of their science experiences) and for practicing informal technical writing (students use organizational structures and writing conventions). Students learn to communicate their thinking in an organized fashion while engaging in the cognitive processes required to develop concepts and build explanations. Having this developmental record of learning also provides an authentic means for assessing students' progress in both scientific thinking and communication skills.

Developing Writing for Literacy in Science

Using student science notebooks in science instruction provides opportunities to address the CCSS for Writing in Science. Grades 6–8 students "write routinely over extended time frames (time for research, reflection, and revision) and shorter time frames (a single sitting or a day or two) for a range of tasks, purposes, and audiences." In addition to providing a structure for recording and analyzing data, notebooks serve as a reference tool from which students can draw information in order to produce derivative products, that is, more formal science writing pieces that have a specific purpose and format. CCSS focus on three text types that students should be writing in science: argument, informational/explanatory writing, and narrative writing. These text types are used in science notebooks and can be developed into derivative products such as reports, articles, brochures, poster boards, electronic presentations, letters, and so forth. Following is a description of these three text types and examples that may be used with FOSS investigations to help students build scientific literacy.

Engaging in Argument

In science, middle school students make claims in the form of statements or conclusions that answer questions or address problems. CCSS Appendix A describes that for students to use "data in a scientifically acceptable form, students marshal evidence and draw on their understanding of scientific concepts to argue in support of their claims." Applying the literacy skills necessary for this type of writing concurrently supports the development of critical science and engineering practices—most notably, engaging in argument. According to *A Framework for K–12 Science Education*, upon which the Next Generation Science Standards (NGSS) are based, middle school students are expected to construct a convincing argument that supports or refutes claims for explanations about the natural and designed world in these ways.

In FOSS, this type of writing makes students' thinking visible. Both informally in their notebooks and formally on assessments, students use deductive and inductive reasoning to construct and defend their explanations. In this way, students deepen their science understanding and exercise the language functions necessary for higher-level thinking, for example, comparing, synthesizing, evaluating, and justifying. To support students in both oral and written argumentation, use the questions and prompts in the *Investigations Guide* that encourage students to use evidence, models, and theories to support their arguments. In addition, be prepared for those teachable moments that provide the perfect stage for spontaneous scientific debate. Here are some general questions to help students deepen their writing.

- Why do you agree or disagree with _____?
- How would you prove/disprove _____?
- What data did you use to make that conclusion that _____?
- Why was it better that _____?

Here are the ways engaging in written argument are developed in the FOSS investigations and can be extended through formal writing.

Response sheets. The FOSS response sheets give students practice in constructing arguments by providing hypothetical situations where they have to apply what they have learned in order to evaluate a claim. For example, one of the response sheets in the **Planetary Science Course** asks students to respond to three students' explanations for the seasons. Students write a paragraph to each student with the

Engaging in Argument
- *Response sheets*
- *Think questions*
- *I-Checks and surveys/posttests*
- *Persuasive writing*

▶ **CCSS NOTE**
This example supports
CCSS.ELA-Literacy.W.1.

Science-Centered Language Development in Middle School

purpose of changing his or her thinking. In order to refute each claim, students must evaluate the validity of the statements and construct arguments based on evidence from the data they've collected during the investigations and logical reasoning that supports their explanation for what causes seasons.

Think questions. Interactive reading in *FOSS Science Resources* is another opportunity for students to engage in written argumentation. Articles include questions that support reading comprehension and extend student thinking about the science content. Asking students to make a claim and provide evidence to support it encourages the use of language functions necessary for higher-level thinking such as evaluating, applying, and justifying. For example, in *FOSS Science Resources: Planetary Science*, students are asked to respond to the following question: Why do you think there are so few craters on Earth and so many on the Moon? After discussion with their peers, students can hone their argumentation skills by writing an argument that answers the question and is supported by the evidence in the *FOSS Science Resources* book as well as data recorded from their experience making model craters.

I-Checks and surveys/posttests. Like the FOSS response sheets, some test items assess students' ability to make a claim and provide evidence to support it. One way is to provide students with data and have them make a claim based on that data and evidence from their prior investigations. Their argument should use logical reasoning to support their ideas. For example, in **Planetary Science**, students are shown images taken from two different planets. They are told that one has a thick atmosphere and the other has no atmosphere. They are asked which image they think came from a planet with an atmosphere and why. Using the images, they can see evidence of craters, and they can draw on their own experiences as well as knowledge acquired through other sources to piece together a logical argument.

Persuasive writing. Formal writing gives students the opportunity to summarize, explain, apply, and evaluate what they have learned in science. It also provides a purpose and audience that motivate students to produce higher-level writing products. The objective of persuasive writing is to convince the reader that a stated interpretation of data is worthwhile and meaningful. In addition to supporting claims with evidence and using logical argument, the writer also uses persuasive techniques such as a call to action. Students can use their informal notebook entries to form the basis of formal persuasive writing in a variety of formats, such as essays, letters, editorials, advertisements, award nominations, informational pamphlets, and petitions. Animal habitats, energy use, weather patterns, landforms, and water sources are just a few science topics that can generate questions and issues for persuasive writing.

Here is a sample of writing frames that can be used to introduce and scaffold persuasive writing (modified from Gibbons, 2002).

Title: _____

The topic of this discussion is _____.

My opinion (position, conclusion) is _____.

There are <number> reasons why I believe this to be true.

First, _____.

Second, _____.

Finally, _____.

On the other hand, some people think _____.

I have also heard people say _____.

However, my claim is that _____ because _____.

Science-Centered Language Development in Middle School

Informational/Explanatory Writing
- *Writing frames*
- *Recursive cycle*

▶ **CCSS NOTE**
Designing, recording, and following procedures in FOSS courses supports CCSS.ELA-Literacy.RST.6–8.3.

▶ **CCSS NOTE**
This example supports CCSS.ELA-Literacy.W.2.

Informational/Explanatory Writing

Informational and explanatory writing requires students to examine and convey complex ideas and information clearly and accurately through the effective selection, organization, and analysis of content. In middle school science, this includes writing scientific procedures and experiments. Described in CCSS Appendix A, informational/explanatory writing answers the questions, What type? What are the components? What are the properties, functions, and behaviors? How does it work? What is happening? Why? In FOSS, this type of writing takes place informally in science notebooks, where students are recording their questions, plans, procedures, data, and answers to the focus questions. It also supports sense making as students attempt to convey what they know in response to questions and prompts, using language functions such as identifying, comparing and contrasting, explaining cause-and-effect relationships, and sequencing.

As an extension of the notebook entries, students can apply their content knowledge to publish formal products such as letters, definitions, procedures, newspaper and magazine articles, posters, pamphlets, and research reports. Strategies such as the writing process (plan, draft, edit, revise, and share) and writing frames (modeling and guiding the use of topic sentences, transition and sequencing words, examples, explanations, and conclusions) can be used to scaffold and help students develop proficiency in science writing.

Writing frames. Here are samples of writing frames (modified from Wellington and Osborne, 2001).

Description

Title: _____

(Identify) The part of the _____ I am describing is the _____.

(Describe) It consists of _____.

(Explain) The function of these parts is _____.

(Example) This drawing shows _____.

Explanation

Title: _____

I want to explain why (how) _____.

An important reason for why (how) this happens is that _____.

Another reason is that _____.

I know this because _____.

Recursive cycle. An effective method for extending students' science learning through writing is the recursive cycle of research (Bereiter, 2002). This strategy emphasizes writing as a process for learning, similar to the way students learn during the active science investigations.

1. Decide on a problem or question to write about.
2. Formulate an idea or a conjecture about the problem or question.
3. Identify a remedy or an answer, and develop a coherent discussion.
4. Gather information (from an experiment, science notebooks, *FOSS Science Resources*, FOSSweb multimedia, books, Internet, interviews, videos, etc.).
5. Reevaluate the problem or question based on what has been learned.
6. Revise the idea or conjecture.
7. Make presentations (reports, posters, electronic presentations, etc.).
8. Identify new needs, and make new plans.

This process can continue for as long as new ideas and questions occur, or students can present a final product in any of the suggested formats.

Narrative Writing

Narrative writing conveys an experience to the reader, usually with sensory detail and a sequence of events. In middle school science, students learn the importance of writing narrative descriptions of their procedures with enough detail and precision to allow others to replicate the experiment. Science also provides a broad landscape of stimulating material for stories, songs, biographies, autobiographies, poems, and plays. Students can enrich their science learning by using organisms or objects as characters; describing habitats and environments as settings; and writing scripts portraying various systems, such as weather patterns, states of matter, and the water, rock, or life cycle.

▶ **CCSS NOTE**
This example supports CCSS.ELA-Literacy.W.7.

▶ **NOTE**
Human characteristics should not be given to organisms (anthropomorphism) in science investigations, only in literacy extensions.

▶ **CCSS NOTE**
This example supports CCSS.ELA-Literacy.W.3.

Science-Centered Language Development in Middle School

> **CCSS NOTE**
> The use of *FOSS Science Resources* supports CCSS.ELA-Literacy.RST.6–8.10.

READING DOMAIN

Reading is an integral part of science learning. Just as scientists spend a significant amount of their time reading each other's published works, students need to learn to read scientific text—to read effectively for understanding, with a critical focus on the ideas being presented.

The articles in *FOSS Science Resources* facilitate sense making as students make connections to the science concepts introduced and explored during the active investigations. Concept development is most effective when students are allowed to experience organisms, objects, and phenomena firsthand before engaging the concepts in text. The text and illustrations help students make connections between what they have experienced concretely and the abstract ideas that explain their observations.

FOSS Science Resources provides students with clear and coherent explanations, ways of visualizing important information, and different perspectives to examine and question. As students apply these strategies, they are, in effect, using some of the same scientific thinking processes that promote critical thinking and problem solving. In addition, the text provides a level of complexity appropriate for middle schoolers to develop high-level reading comprehension skills. This development requires support and guidance as students grapple with more complex dimensions of language meaning, structure, and conventions. To become proficient readers of scientific and other academic texts, students must be armed with an array of reading comprehension strategies and have ample opportunities to practice and extend their learning by reading texts that offer new language, new knowledge, and new modes of thought.

Oral discourse and writing are critical for reading comprehension and for helping students make sense of the active investigations. Use the suggested prompts, questions, and strategies in the *Investigations Guide* to support comprehension as students read from *FOSS Science Resources*. For most of the investigation parts, the articles are designed to follow the active investigation and are interspersed throughout the course. This allows students to acquire the necessary background knowledge in context through active experience before tackling the wider-ranging content and relationships presented in the text. Breakpoints in the readings are suggested in the *Investigations Guide* to support student conceptual development. Some questions make connections between the reading and the student's class experience. Other questions help the students consider the writer's intent. Additional strategies for reading are derived from the seven essential strategies that readers use to help them understand what they read (Keene and Zimmermann, 2007).

> **CCSS NOTE**
> Reading breakpoints in the *Investigations Guide* support CCSS.ELA-Literacy.RST.6–8.8.

C18 Full Option Science System

- Monitor for meaning: discover when you know and when you don't know.
- Use and create schemata: make connections between the novel and the known; activate and apply background knowledge.
- Ask questions: generate questions before, during, and after reading that reach for deeper engagement with the text.
- Determine importance: decide what matters most, what is worth remembering.
- Infer: combine background knowledge with information from the text to predict, conclude, make judgments, and interpret.
- Use sensory and emotional images: create mental images to deepen and stretch meaning.
- Synthesize: create an evolution of meaning by combining understanding with knowledge from other texts/sources.

Reading Comprehension Strategies

Below are some strategies that enhance the reading of expository texts in general and have proven to be particularly helpful in science. Read and analyze the articles beforehand in order to guide students through the text structures and content more effectively.

Build on background knowledge. Activating prior knowledge is critical for helping students make connections between what they already know and new information. Reading comprehension improves when students have the opportunity to think, discuss, and write about what they know about a topic before reading. Review what students learned from the active investigation, provide prompts for making connections, and ask questions to help students recall past experiences and previous exposure to concepts related to the reading.

Create an anticipation guide. Create true-or-false statements related to the key ideas in the reading selection. Ask students to indicate if they agree or disagree with each statement before reading, then have them read the text, looking for the information that supports their true-or-false claims. Anticipation guides connect students to prior knowledge, engage them with the topic, and encourage them to explore their own thinking. To provide a challenge for advanced students, have them come up with the statements for the class.

Draw attention to vocabulary. Check the article for bold faced words students may not know. Review the science words that are already defined in students' notebooks. For new science and nonscience

Reading Comprehension Strategies
- *Build on background knowledge*
- *Create an anticipation guide*
- *Draw attention to vocabulary*
- *Preview the text*
- *Turn and talk*
- *Jigsaw text reading*
- *Note making*
- *Summarize and synthesize*
- *3-2-1*
- *Write reflections*
- *Preview and predict*
- *SQ3R*

▶ **CCSS NOTE**
The example of reviewing what students learned from the active investigation supports CCSS.ELA-Literacy.RST.6–8.9.

▶ **CCSS NOTE**
This example supports CCSS.ELA-Literacy.RST.6–8.4.

Science-Centered Language Development in Middle School

vocabulary words that appear in the reading, have students predict their meanings before reading. During the reading, have students use strategies such as context clues and word structure to see if their predictions were correct. This strategy activates prior knowledge and engages students by encouraging analytical participation with the text.

Preview the text. Give students time to skim through the selection, noting subheads, before reading thoroughly. Point out the particular structure of the text and what discourse markers to look for. For example, most *FOSS Science Resources* articles are written as cause and effect, problem and solution, question and answer, comparison and contrast, description, or sequence. Students will have an easier time making sense of the text if they know what text structure to look for. Model and have students practice analyzing these different types of expository text structures by looking for examples, patterns, and discourse markers. For example, let's look at a passage from *FOSS Science Resources: Planetary Science*.

> An eclipse of the Moon occurs when Earth passes exactly between the Moon and the Sun. [cause and effect] The Moon moves into Earth's shadow during a lunar eclipse. At the time of a full lunar eclipse, Earth's shadow completely covers the disk of the Moon. [description] This is how Earth, the Moon, and the Sun are aligned for a lunar eclipse to be observed. [photograph] Why don't we see a lunar eclipse every month? [question and answer] Because of the tilt of the Moon's orbit around the Earth, Earth's shadow does not fall on the Moon in most months.

Point out how the text in *FOSS Science Resources* is organized (titles, headings, subheadings, questions, and summaries) and if necessary, review how to use the table of contents, glossary, and index. Explain how to scan for formatting features that provide key information (such as boldface type and italics, captions, and framed text) and graphic features (such as tables, graphs, photographs, maps, diagrams, and charts) that help clarify, elaborate, and explain important information in the reading.

While students preview the article, have them focus on the questions that appear in the text, as well as questions at the end of the article. Encourage students to write down questions they have that they think the article will answer.

Turn and talk. When reading as a whole class, stop at key points and have students share their thinking about the selection with the student sitting next to them or with their collaborative group. This strategy helps students process the information and allows everyone to participate in the discussion. When reading in pairs, encourage

▶ **NOTE**
Discourse markers are words or phrases that relate one idea to another. Examples are *however, on the other hand,* and *second.*

▶ **CCSS NOTE**
This example supports CCSS.ELA-Literacy.RST.6-8.5 and CCSS.ELA-Literacy.RST.6-8.6.

▶ **CCSS NOTE**
This example supports CCSS.ELA-Literacy.RST.6-8.7.

students to stop and discuss with their partners. One way to encourage engagement and understanding during paired reading is to have students take turns reading aloud a paragraph or section on a certain topic. The one who is listening then summarizes the meaning conveyed in the passage.

Jigsaw text reading. Students work together in small groups (expert teams) to develop a collective understanding of a text. Each expert team is responsible for one portion of the assigned text. The teams read and discuss their portions to gain a solid understanding of the key concepts. They might use graphic organizers to refine and organize the information. Each expert team then presents its piece to the rest of the class. Or form new jigsaw groups that consist of at least one representative from each expert team. Each student shares with the jigsaw group what their team learned from their particular portion of the text. Together, the participants in the jigsaw group fit their individual pieces together to create a complete picture of the content in the article.

Note making. The more students interact with a reading, the better their understanding. Encourage students to become active readers by asking them to make notes as they read. Studies have shown that note making—especially paraphrasing and summarizing—is one of the most effective means for understanding text (Graham and Herbert, 2010; Applebee, 1984). Some investigation parts include notebook sheets that match pages in *FOSS Science Resources*. This allows students to highlight and underline important points, add notes in the margins, and circle words they do not know. Students can also annotate the article by writing thoughts and questions on self-stick notes. Using symbols or codes can help facilitate comprehension monitoring. Here are some possible symbols students can use to communicate their thinking as they interact with text. (Harvey, 1998).

- * interesting
- BK background knowledge
- ? question
- C confusing
- I important
- L learning something new
- W wondering
- S surprising

> **CCSS NOTE**
> This example supports CCSS.ELA-Literacy.RST.6–8.10.

Science-Centered Language Development in Middle School

Students can also use a different set of symbols while making notes about connections: the readings in *FOSS Science Resources* incorporate the active learning that students gain from the investigations, so that they can make authentic text-to-self (T-S) connections. In other words, what they read reminds them of firsthand experiences, making the article more engaging and easier to understand. Text-to-text (T-T) connections are notes students make when they discover a new idea that reminds them of something they've read previously in another text. Text-to-world (T-W) connections involve the text and more global everyday connections to students' lives.

▶ **CCSS NOTE**
This example supports CCSS.ELA-Literacy.RST.6–8.1 and CCSS.ELA-Literacy.RST.6–8.2.

You can model note-making strategies by displaying a selection of text, using a projection system, a document camera, or an interactive whiteboard. As you read the text aloud, model how to write comments on self-stick notes, and use a graphic organizer in a notebook to enhance understanding.

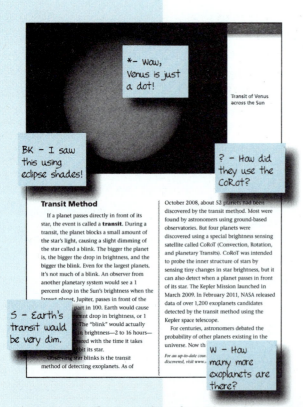

An example of annotated text from *FOSS Science Resources: Planetary Science*

Graphic organizers help students focus on extracting the important information from the reading and analyzing relationships between concepts. This can be done by simply having students make columns in their notebooks to record information and their thinking (Harvey and Goudvis, 2007). Here are two examples.

Notes	Thinking

Facts	Questions	Responses

▶ **CCSS NOTE**
This example supports CCSS.ELA-Literacy.RST.6–8.1 and CCSS.ELA-Literacy.RST.6–8.2.

Summarize and synthesize. Model how to pick out the important parts of the reading selection. Paraphrasing is one way to summarize. Have students write summaries of the reading, using their own words. To scaffold the learning, use graphic organizers to compare and contrast, group, sequence, and show cause and effect. Another method is to have students make two columns in their notebooks. In one column, they record what is important, and in the other, they record their personal responses (what the reading makes them think about). When writing summaries, tell students,

- *Pick out the important ideas.*
- *Restate the main ideas in your own words.*
- *Keep it brief.*

3-2-1. This graphic-organizer strategy gives students the opportunity to synthesize information and formulate questions they still have regarding the concepts covered in an article. In their notebooks, students write three new things they learned, two interesting things worth remembering and sharing, and one question that occurred to them while reading the article. Other options might include three facts, two interesting ideas, and one insight about themselves as learners; three key words, two new ideas, and one thing to think about (modified from Black Hills Special Services Cooperative, 2006).

Write reflections. After reading, ask students to review their notes in their notebooks to make any additions, revisions, or corrections to what they recorded during the reading. This review can be facilitated by using a line of learning. Students draw a line under their original conclusion or under their answer to a question posed at the end of an article. They add any new information as a new narrative entry. The line of learning indicates that what follows represents a change of thinking.

Science-Centered Language Development in Middle School

Preview and predict. Instruct students to independently preview the article, directing attention to the illustrations, photos, boldfaced words, captions, and anything else that draws their attention. Working with a partner, students discuss and write three things they think they will learn from the article. Have partners verbally share their list with another pair of students. The group of four can collaborate to generate one list. Groups report their ideas, and together you create a class list on chart paper.

Read the article aloud, or have students read with a partner aloud or silently. Referring to the preview/prediction list, discuss what students learned. Have them record the most important thing they learned from the reading for comparison with the predictions.

SQ3R. Survey, Question, Read, Recall, Reflect strategy provides an overall structure for before, during, and after reading. Students begin by surveying or previewing the text, looking for features that will help them make predictions about the content. Based on their surveys, students develop questions to answer as they read. They read the selections looking for answers to their questions. Next, they recall what they have learned by retelling a partner and/or recording what they've learned. Finally, they reflect on what they have learned, check to see that they've answered their questions sufficiently, and add any new ideas. Below is a chart students can use to record the SQ3R process in their notebooks.

S Survey	Q Question	R Read	R Recall	R Reflect
Scan the text and record important information.	Ask questions about the subject and what you already know.	Record answers to your questions after you read.	Retell what you learned in your own words.	Did you answer your questions? Record new ideas and comments.

Struggling Readers

For students reading below grade level, the strategies listed on the previous pages can be modified to support reading comprehension by integrating scaffolding strategies such as read-alouds and guided reading. Breaking the reading down into smaller chunks, providing graphic organizers, and modeling reading comprehension strategies can also help students who may be struggling with the text. For additional strategies for English learners, see the supported-reading strategy in the English-Language Development section of this chapter.

Interactive reading aloud. Reading aloud is an effective strategy for enhancing text comprehension. It offers opportunities to model specific reading comprehension strategies and allows students to concentrate on making sense of the content. When modeling, share the thinking processes used to understand the reading (questioning, visualizing, comparing, inferring, summarizing, etc.), then have students share what they observed you thinking about as an active reader.

Guided reading. While the rest of the class is reading independently or in small groups, pull a group aside for a guided reading session. Before reading, review vocabulary words from the investigation and ask questions to activate prior knowledge. Have students preview the text to make predictions, ask questions, and think about text structure. Review reading comprehension strategies they will need to use (monitoring for understanding, asking questions, summarizing, synthesizing, etc.). As students read independently, provide support where needed. Ask questions and provide prompts to guide comprehension. (See the list below for additional strategies.) After reading, have students reflect on what strategies they used to help them understand the text and make connections to the investigation.

- While reading, look for answers to questions and confirm predictions.
- Study graphics, such as pictures, graphs, and tables.
- Reread captions associated with pictures, graphs, and tables.
- Note all italicized and boldfaced words or phrases.
- Reduce reading speed for difficult passages.
- Stop and reread parts that are not clear.
- Read only a section at a time, and summarize after each section.

Struggling Readers
- *Interactive reading aloud*
- *Guided reading*

Science-Centered Language Development in Middle School

SCIENCE-VOCABULARY DEVELOPMENT

Words play two critically important functions in science. First and most important, we play with ideas in our minds, using words. We present ourselves with propositions—possibilities, questions, potential relationships, implications for action, and so on. The process of sorting out these thoughts involves a lot of internal conversation, internal argument, weighing options, and complex linguistic decisions. Once our minds are made up, communicating that decision, conclusion, or explanation in writing or through verbal discourse requires the same command of the vocabulary. Words represent intelligence; acquiring the precise vocabulary and the associated meanings is key to successful scientific thinking and communication.

The words introduced in FOSS investigations represent or relate to fundamental science concepts and should be taught in the context of the investigation. Many of the terms are abstract and are critical to developing science content knowledge and scientific and engineering practices. The goal is for students to use science vocabulary in ways that demonstrate understanding of the concepts the words represent—not to merely recite scripted definitions. The most effective strategies for science-vocabulary development help students make connections to what they already know. These strategies focus on giving new words conceptual meaning through experience; distinguishing between informal, everyday language and academic language; and using the words in meaningful contexts.

Building Conceptual Meaning through Experience

In most instances, students should be presented with new words when they need to know them in the context of the active experience. Words such as *kinetic energy, atmospheric pressure, chemical reaction, photosynthesis,* and *transpiration* are abstract and conceptually loaded. Students will have a much better chance of understanding, assimilating, and remembering the new word (or new meaning) if they can connect it with a concrete experience.

The vocabulary icon appears in the sidebar when students are prompted to record new words in their notebook. The words that appear in bold are critical to understanding the concepts or scientific practices students are learning and applying in the investigation.

When you introduce a new word, students should

- Hear it: students listen as you model the correct contextual use and pronunciation of the word;
- See it: students see the new word written out;
- Say it: students use the new word when discussing their observations and inferences; and
- Write it: students use the new words in context when they write in their notebooks.

Bridging Informal Language to Science Vocabulary

FOSS investigations are designed to tap into students' inquisitive nature and their excitement of discovery in order to encourage lively discussions as they explore materials in creative ways. There should be a lot of talking during the investigations! Your role is to help students connect informal language to the vocabulary used to express specific science concepts. As you circulate during active investigations, you continually model the use of science vocabulary. For example, as students are examining a leaf under the microscope, they will say, "I can see little mouths." You might respond, "Yes, those mouthlike openings are called stomates. They are pores that open and close." Below are some strategies for validating students' conversational language while developing their familiarity with and appreciation for science vocabulary.

Cognitive-content dictionaries. Choose a term that is critical for conceptual understanding of the science investigation. Have students write the term, predict its meaning, write the final meaning after class discussion (using primary language or an illustration), and use the term in a sentence.

Cognitive-Content Dictionary	
New term	kinetic energy
Prediction (clues)	something that moves a lot
Final meaning	motion energy
How I would use it in a sentence	Fast-moving particles have more kinetic energy than slow-moving particles.

Bridging Informal Language to Science Vocabulary
- *Cognitive-content dictionaries*
- *Concept maps*
- *Semantic webs*
- *Word associations*
- *Word sorts*

Science-Centered Language Development in Middle School

Concept maps. Select six to ten related science words. Have students write them on self-stick notes or cards. Have small groups discuss how the words are related. Students organize words in groups and glue them down or copy them on a sheet of paper. Students draw lines between the related words. On the lines, they write words describing or explaining how the concept words are related.

Semantic webs. Select a vocabulary word, and write it in the center of a piece of paper (or on the board for the whole class). Brainstorm a list of words or ideas that are related to the first word. Group the words and concepts into several categories, and attach them to the central word with lines, forming a web (modified from Hamilton, 2002).

Word associations. In this brainstorming activity, you say a word, and students respond by writing the first word that comes to mind. Then students share their words with the class. This activity builds connections to students' prior frames of reference.

Word sorts. Have students work with a partner to make a set of word cards using new words from the investigation. Have them group the words in different ways, for example, synonyms, root words, and conceptual connections.

Using Science Vocabulary in Context

For a new vocabulary word to become part of a student's functional vocabulary, he or she must have ample opportunities to hear and use it. Vocabulary terms are used in the activities through teacher talk, whole-class and small-group discussions, writing in science notebooks, readings, and assessments. Other methods can also be used to reinforce important vocabulary words and phrases.

Word wall. Use chart paper to record science content and procedural words as they come up during and after the investigations. Students will use this word wall as a reference.

Drawings and diagrams. For English learners and visual learners, use a diagram to review and explain abstract content. Ahead of time, draw an illustration lightly, almost invisibly, with pencil on chart paper. You can do this easily by projecting the image onto the paper. When it's time for the investigation, trace the illustration with markers as you introduce the words and phrases to students. Students will be amazed by your artistic ability.

Science Vocabulary Strategies
- *Word wall*
- *Drawings and diagrams*
- *Cloze activity*
- *Word wizard*
- *Word analysis/word parts*
- *Breaking apart words*
- *Possible sentences*
- *Reading*
- *Glossary*
- *Index*
- *Poems, chants, and songs*

Cloze activity. Structure sentences for students to complete, leaving out the vocabulary words. This can be done as a warm-up with the words from the previous day's lesson. Here's an example from the **Earth History Course**.

> Teacher: *The removal and transportation of loose earth materials is called _____.*
>
> Students: *Erosion.*

Word wizard. Tell students that you are going to lead a word activity. You will be thinking of a science vocabulary word. The goal is to figure out the word. Provide hints that have to do with parts of a definition, root word, prefix, suffix, and other relevant components. Students work in teams of two to four. Provide one hint, and give teams 1 minute to discuss. One team member writes the word on a piece of paper or on the whiteboard, using dark marking pens. Each team holds up its word for only you to see. After the third clue, reveal the word, and move on to the next word. Here's an example.

1. *Part of the word means green.*
2. *They are found in plant cells.*
3. *They look like tiny green spheres or ovals.*

The word is **chloroplasts**.

Word analysis/word parts. Learning clusters of words that share a common origin can help students understand content-area texts and connect new words to familiar ones. Here's an example: *geology, geologist, geological, geography, geometry, geophysical*. This type of contextualized teaching meets the immediate need of understanding an unknown word while building generative knowledge that supports students in figuring out difficult words for future reading.

Breaking apart words. Have teams of two to four students break a word into prefix, root word, and suffix. Give each team different words, and have each team share the parsed elements of the word with the whole class. Here's an example.

> *photosynthesis*
>
> Prefix = *photo*: meaning light
>
> Root = *synthesis*: meaning to put together

Science-Centered Language Development in Middle School

Possible sentences. Here is a simple strategy for teaching word meanings and generating class discussion.

1. Choose six to eight key concept words from the text of an article in *FOSS Science Resources*.

2. Choose four to six additional words that students are more likely to know something about.

3. Put the list of ten to fourteen words on the board or project it. Provide brief definitions as needed.

4. Ask students to devise sentences that include two or more words from the list.

5. On chart paper, write all sentences that students generate, both coherent and otherwise.

6. Have students read the article from which the words were extracted.

7. Revisit students' sentences, and discuss whether the sentences are sensible based on the passage or how they could be modified to be more coherent.

Reading. After the active investigation, students continue to develop their understanding of the vocabulary words and the concepts those words represent by listening to you read aloud, reading with a partner, or reading independently. Use strategies discussed in the Reading Domain section to encourage students to articulate their thoughts and practice the new vocabulary.

Glossary. Emphasize the vocabulary words students should be using when they answer the focus question in their science notebooks. The glossary in *FOSS Science Resources* or on FOSSweb can be used as a reference.

Index. Have students create an index at the back of their notebooks. There they can record new vocabulary words and the notebook page where they defined and used the new words for the first time in the context of the investigation.

Poems, chants, and songs. As extensions or homework assignments, ask students to create poems, raps, chants, or songs, using vocabulary words from the investigation.

> **NOTE**
> See the Science Notebooks in Middle School chapter for an example of an index.

ENGLISH-LANGUAGE DEVELOPMENT

Active investigations, together with ample opportunities to develop and use language, provide an optimal learning environment for English learners. This section highlights opportunities for English-language development (ELD) in FOSS investigations and suggests other best practices for facilitating both the learning of new science concepts and the development of academic literacy. For example, the hands-on structure of FOSS investigations is essential for the conceptual development of science content knowledge and the habits of mind that guide and define scientific and engineering practices. Students are engaged in concrete experiences that are meaningful and that provide a shared context for developing understanding—critical components for language acquisition.

When getting ready for an investigation, review the steps and determine the points where English learners may require scaffolds and where the whole class might benefit from additional language-development supports. One way to plan for ELD integration in science is to keep in mind four key areas: prior knowledge, comprehensible input, academic language development, and oral practice. The ELD chart lists examples of universal strategies for each of these components that work particularly well in teaching science.

▶ **NOTE**
English-language development refers to the advancement of students' ability to read, write, and speak English.

English-Language Development (ELD)	
Activating prior knowledge	**Using comprehensible input**
• Inquiry chart • Circle map • Observation poster • Quick write • Kit inventory	• Content objectives • Multiple exposures • Visual aids • Supported reading • Procedural vocabulary
Developing academic language	**Providing oral practice**
• Language objectives • Sentence frames • Word wall, word cards, drawings • Concept maps • Cognitive content dictionaries	• Small-group discussions • Science talk • Oral presentations • Poems, chants, and songs • Teacher feedback

Science-Centered Language Development in Middle School

Science-Centered Language Development in Middle School

> **NOTE**
> Language forms and structures are the internal grammatical structure of words and how those words go together to make sentences.

Students acquiring English benefit from scaffolds that support the language forms and functions necessary for the academic demands of the science course, that is, accessing science text, participating in productive oral discourse, and engaging in science writing. The table at the end of this section (starting on page 38) provides a resource to help students organize their thinking and structure their speaking and writing in the context of the science and engineering practices. The table identifies key language functions exercised during FOSS investigations and provides examples of sentence frames students can use as scaffolds.

For example, if students are planning an investigation to learn more about insect structures and behaviors, the language objective might be "Students plan and design an investigation that answers a question about the hissing cockroach's behavior." For students who need support, a sentence frame that prompts them to identify the variables in the investigation would provide language forms and structures appropriate for planning their investigation. As a scaffold, sentence frames can also help them write detailed narratives of their procedure. Here's an example from the table.

> **NOTE**
> The complete table appears at the end of the English-Language Development section starting on page 38.

Language functions	Language objectives	Sentence frames
Planning and carrying out investigations		
Design Sequence Strategize Evaluate	Plan controlled experiments with multiple trials. Identify independent variable and dependent variable. Discuss, describe, and evaluate the methods for collecting data.	To find out _____, I will change _____. I will not change _____. I will measure _____. I will observe _____. I will record the data by _____. First, I will _____, and then I will _____. To learn more about _____, I will need _____ to _____.

Full Option Science System

Activating Prior Knowledge

When an investigation engages a new concept, students first recall and discuss familiar situations, objects, or experiences that relate to and establish a foundation for building new knowledge and conceptual understanding. Eliciting prior knowledge also supports learning by motivating interest, acknowledging culture and values, and checking for misconceptions and prerequisite knowledge. This is usually done in the first steps of Guiding the Investigation in the form of a discussion, presentation of new materials, or a written response to a prompt. The tools outlined below can also be used before beginning an investigation to establish a familiar context for launching into new material.

Activating Prior Knowledge
- *Circle maps*
- *Observation posters*
- *Quick writes*
- *Kit inventories*

Circle maps. Draw two concentric circles on chart paper. In the smaller circle, write the topic to be explored. In the larger circle, record what students already know about the subject. Ask students to think about how they know or learned what they already know about the topic. Record the responses outside the circles. Students can also do this independently in their science notebooks.

An example of a circle map

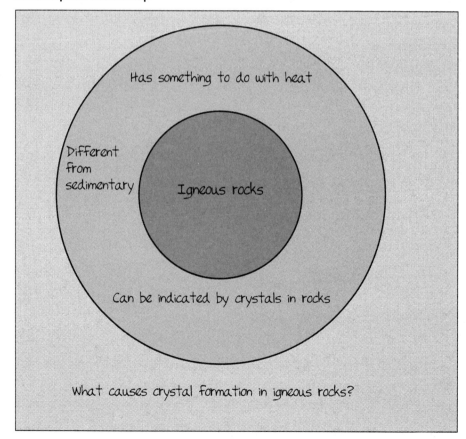

Science-Centered Language Development in Middle School

Observation posters. Make observation posters by gluing or taping pictures and artifacts relevant to the module or a particular investigation onto pieces of blank chart paper or poster paper. Hang them on the wall in the classroom, and have students rotate in small groups to each poster. At each poster, students discuss their observations with their partners or small groups and then record (write or draw) an observation, a question, a prediction, or an inference about the pictures as a contribution to the commentary on the poster.

As a variation on this strategy, give a set of pictures to each group to pass around. Have them choose one and write what they notice, what they infer, and questions they have in their notebooks.

Quick writes. Ask students what they know about the topic of the investigation. Responses can be recorded independently as a quick write in notebooks and then shared collaboratively. Do not correct misconceptions initially. Periodically revisit the quick-write ideas as a whole class, or have students review their notebook entries to correct, confirm, or complete their original thoughts as they acquire new information (possibly using a line of learning). At the conclusion of the investigation, students should be able to express mastery of the new conceptual material.

Kit inventories. Introduce each item from the FOSS kit used in the investigation, and ask students questions to get them thinking about what each item is and where they may have seen it before. Have them describe the objects and predict how they will be used in the investigation.

Comprehending Input

To initiate their own sense making, students must be able to access the information presented to them. We refer to this ability as comprehending input. Students must understand the essence of new ideas and concepts before beginning to construct new scientific meaning. The strategies for comprehensible input used in FOSS ensure that the instruction is understandable while providing students with the opportunity to grapple with new ideas and the critically important relationships between concepts. Additional tools such as repetition, visual aids, emphasis on procedural vocabulary, and auditory reinforcement can also be used to enhance comprehensible input for English learners.

Comprehending Input
- *Content objectives*
- *Multiple exposures*
- *Visual aids*
- *Supported reading*
- *Procedural vocabulary*

Content objectives. The focus question for each investigation part frames the activity objectives—what students should know or be able to do at the end of the part. Making the learning objectives clear and explicit prepares English learners to process the delivery of new information, and helps you maintain the focus of the investigation. Write the focus question on the board, have students read it aloud and transcribe it into their science notebooks, and have students answer the focus question at the end of the investigation part. You then check their responses for understanding.

Multiple exposures. Repeat an activity in an analogous but slightly different context, ideally one that incorporates elements that are culturally relevant to students. For example, as a homework assignment for landforms, have students interview their parents about landforms common in the area of their ancestry.

Visual aids. On the board or chart paper, write out the steps for conducting the investigation. This provides a visual reference. Include illustrations if necessary. Use graphic representations (illustrations drawn and labeled in front of students) to review the concepts explored in the active investigations. In addition to the concrete objects in the kit, use realia to augment the activity, to help English learners build understanding and make cultural connections. Graphic organizers (webs, Venn diagrams, T-tables, flowcharts, etc.) aid comprehension by helping students see how concepts are related.

Supported reading. In addition to the reading comprehension strategies suggested in the Reading Domain section of this chapter, English learners can also benefit from methods such as front-loading key words, phrases, and complex text structures before reading; using

Science-Centered Language Development in Middle School

Science-Centered Language Development in Middle School

Procedural Vocabulary
Add
Analyze
Assemble
Attach
Calculate
Change
Classify
Collect
Communicate
Compare
Connect
Construct
Contrast
Demonstrate
Describe
Determine
Draw
Evaluate
Examine
Explain
Explore
Fill
Graph
Identify
Illustrate
Immerse
Investigate
Label
List
Measure
Mix
Observe
Open
Order
Organize
Pour
Predict
Prepare
Record
Represent
Scratch
Separate
Sort
Stir
Subtract
Summarize
Test
Weigh

preview-review (main ideas are previewed in the primary language, read in English, and reviewed in the primary language); and having students use sentence frames specifically tailored to record key information and/or graphic organizers that make the content and the relationship between concepts visually explicit from the text as they read.

Procedural vocabulary. Make sure students understand the meaning of the words used in the directions for an investigation. These may or may not be science-specific words. Use techniques such as modeling, demonstrating, and body language (gestures) to explain procedural meaning in the context of the investigation. The words students will encounter in FOSS include those listed in the sidebar. To build academic literacy, English learners need to learn the multiple meanings of these words and their specific meanings in the context of science.

Developing Academic Language

As students learn the nuances of the English language, it is critical that they build proficiency in academic language in order to participate fully in the cognitive demands of school. *Academic language* refers to the more abstract, complex, and specific aspects of language, such as the words, grammatical structure, and discourse markers that are needed for higher cognitive learning. FOSS investigations introduce and provide opportunities for students to practice using the academic vocabulary needed to access and engage with science ideas.

Language objectives. Consider the language needs of English learners and incorporate specific language-development objectives that will support learning the science content of the investigation, such as a specific way to expand use of vocabulary by looking at root words, prefixes, and suffixes; a linguistic pattern or structure for oral discussion and writing; or a reading comprehension strategy. Recording in science notebooks is a productive way to optimize science learning and language objectives. For example, in the **Earth History Course**, one language objective might be "Students will apply techniques for rock observations to compare and contrast sedimentary and igneous rocks. They will discuss and record their observations in their notebooks in an organized manner."

Vocabulary development. The Science-Vocabulary Development section in this chapter describes the ways science vocabulary is introduced and developed in the context of an active investigation and suggests methods and strategies that can be used to support vocabulary development during science instruction. In addition to science vocabulary, students need to learn the nonscience vocabulary that facilitates deeper understanding and communication skills. Words such as *release, convert, beneficial, produce, receive, source,* and *reflect* are used in the investigations and *FOSS Science Resources* and are frequently used in other content areas. Learning these academic-vocabulary words gives students a more precise and complex way of practicing and communicating productive thinking. Consider using the strategies described in the Science-Vocabulary Development section to explicitly teach targeted, high-leverage words that can be used in multiple ways and that can help students make connections to other words and concepts. Sentence frames, word walls, concept maps, and cognitive-content dictionaries are strategies that have been found to be effective with academic-vocabulary development.

Science-Centered Language Development in Middle School

Scaffolds That Support Science and Engineering Practices

Language functions	Language objectives	Sentence frames
Asking questions and defining problems		
Inquire Define a problem	Ask questions to solicit information about phenomena, models, or unexpected results; determine the constraints and criteria of a problem.	I wonder why _____ . What happens when _____? What if _____? What does _____? What can _____? What would happen if _____? How does _____ affect _____? How can I find out if _____? Which _____ is better for _____?
Planning and carrying out investigations		
Design Sequence Strategize Evaluate	Plan controlled experiments with multiple trials. Identify independent variable and dependent variable. Discuss, describe, and evaluate the methods for collecting data.	To find out _____, I will change _____. I will not change _____. I will measure _____. I will observe _____. I will record the data by _____. First, I will _____, and then I will _____. To learn more about _____, I will need _____ to _____.

Language functions	Language objectives	Sentence frames
Planning and carrying out investigations *(continued)*		
Describe	Write narratives using details to record sensory observations and connections to prior knowledge.	I observed/noticed ____. When I touch the ____, I feel ____. It smells ____. It sounds ____. It reminds me of ____, because ____.
Organize Compare Classify	Make charts and tables: use a T-table or chart for recording and displaying data.	The table compares ____ and ____
Sequence Compare	Record changes over time, and describe cause-and-effect relationships.	At first, ____, but now ____. We saw that first ____, then ____, and finally ____. When I ____, it ____. After I ____, it ____.
Draw Label Identify	Draw accurate and detailed representations; identify and label parts of a system using science vocabulary, with attention to form, location, color, size, and scale.	The diagram shows ____. ____ is shown here. ____ is ____ times bigger than ____. ____ is ____ times smaller than ____.
Analyzing and interpreting data		
Enumerate Compare Represent	Use measures of variability to analyze and characterize data; decide when and how to use bar graphs, line plots, and two-coordinate graphs to organize data.	The mean is ____. The median is ____. The mode is ____. The range is ____. The x-axis represents ____ and the y-axis represents ____. The units are expressed in ____.

Science-Centered Language Development in Middle School

Science-Centered Language Development in Middle School

Language functions	Language objectives	Sentence frames
Analyzing and interpreting data *(continued)*		
Compare Classify Sequence	Use graphic organizers and narratives to express similarities and differences, to assign an object or action to the category or type to which it belongs, and to show sequencing and order.	This _____ is similar to _____ because _____. This _____ is different from _____ because _____. All these are _____ because _____. _____, _____, and _____ all have/are _____.
Analyze	Use graphic organizers, narratives, or concept maps to identify part/whole or cause-and-effect relationships. Express data in qualitative terms such as more/fewer, higher/lower, nearer/farther, longer/shorter, and increase/decrease; and quantitatively in actual numbers or percentages.	The _____ consists of _____. The _____ contains _____. As _____, then _____. When I changed _____, then _____ happened. The more/less _____, then _____.
Developing and using models		
Represent Predict Explain	Construct and revise models to predict, represent, and explain.	If _____, then _____, therefore _____. The _____ represents _____. _____ shows how _____. You can explain _____ by _____.

Language functions	Language objectives	Sentence frames
Using mathematics and computational thinking		
Symbolize Measure Enumerate Estimate	Use mathematical concepts to analyze data.	The ratio of ____ is ____ to ____. The average is ____. Looking at ____, I think there are ____. My prediction is ____.
Constructing explanations and designing solutions		
Infer Explain	Construct explanations based on evidence from investigations, knowledge, and models; use reasoning to show why the data are adequate for the explanation or conclusion.	I claim that ____. I know this because ____. Based on ____, I think ____. As a result of ____, I think ____. The data show ____, therefore, ____. I think ____ means ____ because ____. I think ____ happened because ____.
Provide evidence	Use qualitative and quantitative data from the investigation as evidence to support claims. Use quantitative expressions using standard metric units of measurement such as cm, mL, °C.	My data show ____. My evidence is ____. The relationship between the variables is ____. The model of ____ shows that ____.

Science-Centered Language Development in Middle School

Science-Centered Language Development in Middle School

Language functions	Language objectives	Sentence frames
Engaging in argument from evidence		
Discuss Persuade Synthesize Negotiate Suggest	Use oral and written arguments supported by evidence and reasoning to support or refute an argument for a phenomenon or a solution to a problem.	I think ___ because___. I agree/disagree with ___ because_____. What you are saying is _____. What do you think about _____? What if _____? I think you should try ___. Another way to interpret the data is _____.
Critique Evaluate Reflect	Evaluate competing design solutions based on criteria; compare two arguments from evidence to identify which is better.	____ makes more sense because ____. ____ is a better design _____ because it ____. Comparing ___ to ___ shows that _____. One discrepancy is ____. ____ is inconsistent with _____. Another way to determine _____ is to _____. I used to think ____, but now I think ____. I have changed my thinking about ____. I am confused about ____ because ____. I wonder ____.
Obtaining, evaluating, and communicating information		
(This practice includes all functions described in the other practices above.)		

Full Option Science System

REFERENCES

Applebee, A. 1984. "Writing and Reasoning." *Review of Educational Research* 54 (Winter): 577–596.

Bereiter, C. 2002. *Education and Mind in the Knowledge Age.* Hillsdale, NJ: Erlbaum.

Black Hills Special Services Cooperative. 2006. "3-2-1 Strategy." In *On Target: More Strategies to Guide Learning.* Rapid City, SD: South Dakota Department of Education.

Gibbons, P. 2002. *Scaffolding Language, Scaffolding Learning.* Portsmouth, NH: Heinemann.

Graham, S., and M. Herbert. 2010. *Writing to Read: Evidence for How Writing Can Improve Reading.* New York: Carnegie.

Hamilton, G. 2002. *Content-Area Reading Strategies: Science.* Portland, ME: Walch Publishing.

Harvey, S. 1998. *Nonfiction Matters: Reading, Writing, and Research in Grades 3–8.* Portland, ME: Stenhouse.

Harvey, S., and A. Goudvis. 2007. *Strategies That Work: Teaching Comprehension for Understanding and Engagement.* Portland, ME: Stenhouse.

Keeley, P. 2008. *Science Formative Assessment: 75 Practical Strategies for Linking Assessment, Instruction, and Learning.* Thousand Oaks, CA: Corwin Press.

Keene, E., and S. Zimmermann. 2007. *Mosaic of Thought: The Power of Comprehension Strategies.* 2nd ed. Portsmouth, NH: Heinemann.

Keys, C. 1999. *Revitalizing Instruction in Scientific Genres: Connecting Knowledge Production with Writing to Learn in Science.* Athens: University of Georgia.

Lieberman, G. A., and L. L. Hoody. 1998. *Closing the Achievement Gap: Using the Environment as an Integrating Context for Learning.* San Diego, CA: State Education and Environment Roundtable.

National Governors Association Center for Best Practices, Council of Chief State School Officers. 2010. *Common Core State Standards for English Language Arts & Literacy in History/Social Studies, Science, and Technical Subjects.* Washington, DC: National Governors Association Center for Best Practices, Council of Chief State School Officers.

National Research Council. 2012. *A Framework for K–12 Science Education: Practices, Crosscutting Concepts, and Core Ideas.* Committee on a Conceptual Framework for New K–12 Science Education Standards. Board on Science Education, Division of Behavioral and Social Sciences and Education. Washington, DC: The National Academies Press.

> **NOTE**
> For additional resources and updated references, go to FOSSweb.

Norris, S. P., and L. M. Phillips. 2003. "How Literacy in Its Fundamental Sense Is Central to Scientific Literacy." *Science Education* 87 (2).

Ostlund, K. 1998. "What the Research Says about Science Process Skills: How Can Teaching Science Process Skills Improve Student Performance in Reading, Language Arts, and Mathematics?" *Electronic Journal of Science Education* 2 (4).

Wellington, J., and J. Osborne. 2001. *Language and Literacy in Science Education*. Buckingham, UK: Open University Press.

Winokur, J., and K. Worth. 2006. "Talk in the Science Classroom: Looking at What Students and Teachers Need to Know and Be Able to Do." In *Linking Science and Literacy in the K–8 Classroom*, ed. R. Douglas, K. Worth, and W. Binder. Arlington, VA: NSTA Press.

FOSSweb and Technology

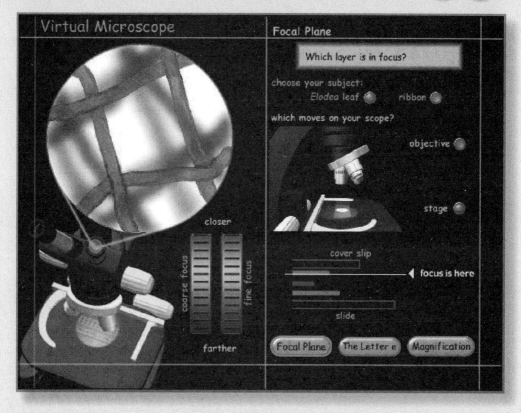

FOSSweb and Technology

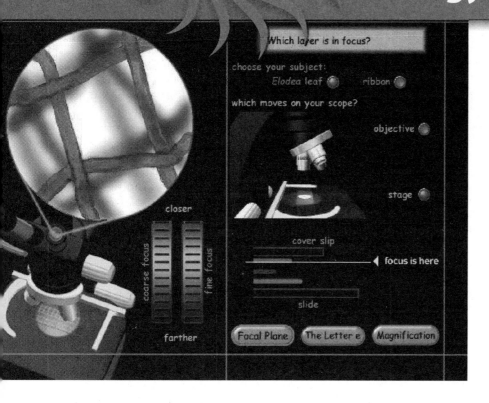

Contents

Introduction D1

Requirements for Accessing
FOSSweb D2

Troubleshooting and
Technical Support D6

INTRODUCTION

FOSSweb technology is an integral part of the **Diversity of Life Course**. It provides students with the opportunity to access and interact with simulations, images, video, and text—digital resources that can enhance their understanding of life science concepts. Different sections of digital resources are incorporated into each investigation during the course. Each use is marked with the technology icon in the *Investigations Guide*. You will sometimes use the digital resources to make presentations to the class. At other times, individuals or small groups of students will work with the digital resources to review concepts or reinforce their understanding.

The FOSSweb components are not optional. To prepare to use these digital resources, you should have at a minimum one device with Internet access that can be displayed to the class by an LCD projector with an interactive whiteboard or a large screen arranged for class viewing. Access to a computer lab or to enough devices in your classroom for students to work in small groups is also required during one investigation, and recommended during others.

The digital resources are available online at www.FOSSweb.com for teachers and students. We recommend you access FOSSweb well in advance of starting the course to set up your teacher-user account and become familiar with the resources.

Full Option Science System   *Copyright © The Regents of the University of California*

FOSSweb and Technology

REQUIREMENTS FOR ACCESSING FOSSWEB

You'll need to have a few things in place on your device before accessing FOSSweb. Once you're online, you'll create a FOSSweb account. All information in this section is updated as needed on FOSSweb.

Creating a FOSSweb Teacher Account

By creating a FOSSweb teacher account, you can personalize FOSSweb for easy access to the courses you are teaching. When you log in, you will be able to add courses to your "My FOSS Modules" area and access Resources by Investigation for the **Diversity of Life Course**. This makes it simple to select the investigation and part you are teaching and view all the digital resources connected to that part.

Students and families can also access course resources through FOSSweb. You can set up a class account and class pages where students will be able to access notes from you about assignments and digital resources.

Setting up an account. Set up an account on FOSSweb so you can access the site when you begin teaching a course. Go to FOSSweb to register for an account—complete registration instructions are available online.

Entering your access code. Once your account is set up, go to FOSSweb and log in. The first time you log in, you will need to enter your access code. Your access code should be printed on the inside cover of your *Investigations Guide*. If you cannot find your FOSSweb access code, contact your school administrator, your district science coordinator, or the purchasing agent for your school or district.

Familiarize yourself with the layout of the site and the additional resources available by using your account login. From your course page, you will be able to access teacher masters, assessment masters, notebook sheets, and other digital resources. Explore the Resources by Investigation section, as this will help you plan. It lists the digital resources, notebook sheets, teacher masters, and readings for each investigation part. There are also a variety of beneficial resources on FOSSweb that can be used to assist with teacher preparation and materials management.

Setting up class pages and student accounts. To enable your students to log in to FOSSweb to access the digital resources and see class assignments, set up a class page and generate a user name and password for the class. To do so, log in to FOSSweb and go to your teacher homepage. Under My Class Pages, follow the instructions to create a new class page and to leave notes for students.

If a class page and student accounts are not set up, students can always access digital resources by visiting FOSSweb.com and choosing to visit the site as a guest.

FOSSweb Technical Requirements

To use FOSSweb, your device must meet minimum system requirements and have a compatible browser and recent versions of any required plug-ins. The system requirements are subject to change. You must visit FOSSweb to review the most recent minimum system requirements.

Preparing your browser. FOSSweb requires a supported operating system with current versions of all required plug-ins. You may need administrator privileges on your device in order to install the required programs and/or help from your school's technology coordinator. Check compatibility for each device you will use to access FOSSweb by accessing the Technical Requirements page on FOSSweb. The information at FOSSweb contains the most up-to-date technical requirements.

https://www.FOSSweb.com/tech-specs-and-info

Support for plug-ins and reader. Any required Adobe plug-ins are available on www.Adobe.com as free downloads. If required, QuickTime is available free of charge from www.Apple.com. FOSS does not support these programs. Please go to the program's website for troubleshooting information.

FOSSweb and Technology

Accessing FOSS Diversity of Life Digital Resources

When you log in to FOSSweb, the most useful way to access course materials on a daily basis is the Resources by Investigation section of the **Diversity of Life Course** page. This section lists the digital resources, student and teacher sheets, readings, and focus questions for each investigation part. Each of these items is linked so you can click and go directly to that item.

Students will access digital resources from the Resource Room, accessible from the class page you've set up. Explore where the activities reside in the Resource Room. At various points in the course, students will access interactive simulations, images, videos, and animations from FOSSweb.

Other Technology Considerations

Firewall or proxy settings. If your school has a firewall or proxy server, contact your IT administrator to add explicit exceptions in your proxy server and firewall for these servers:

- fossweb.com
- fossweb.schoolspecialty.com
- archive.fossweb.com
- science.video.schoolspecialty.com
- a445.w10.akamai.net

Classroom technology setup. FOSS has a number of digital resources and makes every effort to accommodate users with different levels of access to technology. The digital resources can be used in a variety of ways and can be adapted to a number of classroom setups.

Teachers with classroom devices and an LCD projector, an interactive whiteboard, or a large screen will be able to show multimedia to the class. If you have access to a computer lab, or enough devices in your classroom for students to work in small groups, you can set up time for students to use the FOSSweb digital resources during the school day.

Displaying digital content. You might want to digitally display the notebook and teacher masters during class. In the Resources by Investigation section of FOSSweb, you'll have the option of downloading the masters "to project" or "to copy." Choose "to project" if you plan on projecting the masters to the class. These masters are optimized for a projection system and allow you to type into them while they are displayed. The "to copy" versions are sized to minimize paper use when photocopying for the class, and to fit optimally into student notebooks.

If this projection technology is not available to you, consider making transparencies of the notebook and teacher masters for use with an overhead projector when the Getting Ready section indicates a need to project these sheets.

▶ **NOTE**
FOSSweb activities are designed for a minimum screen size of 1024 × 768. It is recommended that you adjust your screen resolution to 1024 × 768 or higher.

FOSSweb and Technology

TROUBLESHOOTING AND TECHNICAL SUPPORT

If you experience trouble with FOSSweb, you can troubleshoot in a variety of ways.

1. Visit FOSSweb and make sure your devices meet the minimum system requirements

    https://www.FOSSweb.com/tech-specs-and-info

2. Check the FAQs on FOSSweb for additional information that may help resolve the problem.

3. Try emptying the cache from your browser and/or quitting and relaunching it.

4. Restart your device, and make sure all hardware turns on and is connected correctly.

5. If your school has a firewall or proxy server, contact your IT administrator to add explicit exceptions in your proxy server and firewall for the servers listed on the previous page in this chapter.

If you are still experiencing problems after taking these steps, send FOSS Technical Support an e-mail at support@FOSSweb.com. In addition to describing the problem, include the following information about your device: name of device, operating system, browser and version, plug-ins and versions. This will help us troubleshoot the problem.

Science Notebook Masters

Living/Nonliving Card Sort

Card name	L	NL	U
Amoeba			
Apple			
Baby			
Blue cheese			
Blue-green algae			
Bread mold			
Cactus			
Clouds			
Coral			
Corn			
Cotton boll			
E. coli			
Eggs			
Fire			
Horse			
Jellyfish			
Kelp			

Card name	L	NL	U
Mushrooms			
Onions			
Potatoes			
Rhinovirus			
Robot			
Rocking horse			
Spider and web			
Streptococcus			
Sulfolobus			
Sun			
Tornado			
Trees and leaves			
Yeast			
Yogurt			

Five-Materials Observation

Liquid number _____

	First observations (dry) (include drawings)	Changes observed after 10 minutes (include drawings)	Changes observed after 24 hours (include drawings)	Changes observed after _____ (include drawings)
A				
B				
C				
D				
E				

Five-Materials Observation

Liquid number _____

	First observations (dry) (include drawings)	Changes observed after 10 minutes (include drawings)	Changes observed after 24 hours (include drawings)	Changes observed after _____ (include drawings)
A				
B				
C				
D				
E				

Life in Different Environments

Liquid 1 _____ Material _____

	What evidence of life do you observe?
A	
B	
C	
D	
E	

Liquid 2 _____ Material _____

	What evidence of life do you observe?
A	
B	
C	
D	
E	

Liquid 3 _____ Material _____

	What evidence of life do you observe?
A	
B	
C	
D	
E	

The Microscope

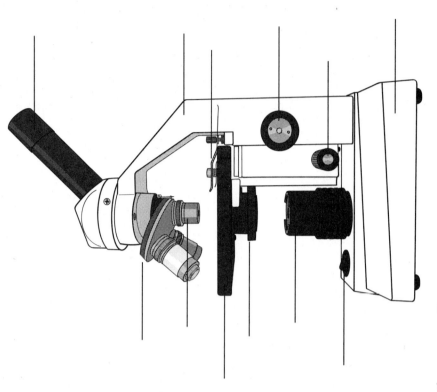

Microscope Care

- Always use two hands to carry a microscope—one hand holding the **arm** and one hand under the **base**. If the microscope has a power cord, gather that up to keep it from getting underfoot.
- Always wipe up any water that is on the scope after a session.
- Always cover the microscope with a dustcover to keep it clean.
- Use *only* lens paper to clean the lenses of a microscope. Do not use paper towels or tissues, as they will scratch the lenses.

The Microscope

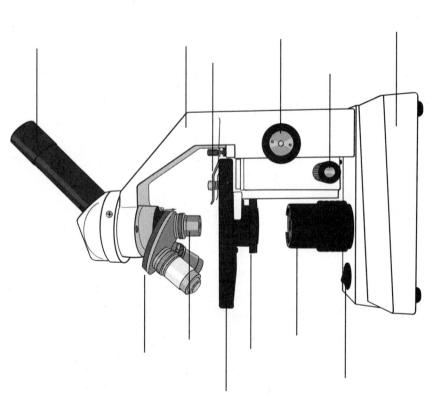

Microscope Care

- Always use two hands to carry a microscope—one hand holding the **arm** and one hand under the **base**. If the microscope has a power cord, gather that up to keep it from getting underfoot.
- Always wipe up any water that is on the scope after a session.
- Always cover the microscope with a dustcover to keep it clean.
- Use *only* lens paper to clean the lenses of a microscope. Do not use paper towels or tissues, as they will scratch the lenses.

Microscope Use and Practice

1. Set up the microscope with the **arm** toward/**arm** away from you.
2. Rotate the **clips** out of the way.
3. Put the prepared slide on the center of the **stage**. The specimen should be placed over the hole where the light comes through.
4. Start with the **low-power (4X) objective**.
5. Use the **coarse focus knob** to bring the objective lens close to the slide. Watch from the side and carefully check to make sure the objective does not touch the slide.
6. Look through the **eyepiece**. Bring the specimen into focus by turning the coarse focus knob so that the objective lens moves away from the stage. (Never use the coarse focus knob to move the lens toward the stage while looking through the eyepiece. You can easily break a slide or damage the lens.)
7. Adjust the amount of light coming to the specimen, using the **diaphragm** located under the stage.
8. Once the specimen is in focus, do not turn either of the focus knobs yet. Rotate the **medium-power (10X) objective** into place and now carefully adjust the focus using the **fine focus knob**. Move to the **high-power (40X) objective** and use the **fine focus knob** to adjust the focus.
9. When you are done,
    a. take the slide off the stage and clean it as directed,
    b. turn off the light,
    c. rotate the objectives to the low-power objective, and
    d. cover the scope with the dustcover and store it as directed.

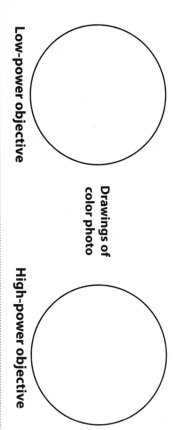

Drawings of color photo

Low-power objective

High-power objective

Microscope Use and Practice

1. Set up the microscope with the **arm** toward/**arm** away from you.
2. Rotate the **clips** out of the way.
3. Put the prepared slide on the center of the **stage**. The specimen should be placed over the hole where the light comes through.
4. Start with the **low-power (4X) objective**.
5. Use the **coarse focus knob** to bring the objective lens close to the slide. Watch from the side and carefully check to make sure the objective does not touch the slide.
6. Look through the **eyepiece**. Bring the specimen into focus by turning the coarse focus knob so that the objective lens moves away from the stage. (Never use the coarse focus knob to move the lens toward the stage while looking through the eyepiece. You can easily break a slide or damage the lens.)
7. Adjust the amount of light coming to the specimen, using the **diaphragm** located under the stage.
8. Once the specimen is in focus, do not turn either of the focus knobs yet. Rotate the **medium-power (10X) objective** into place and now carefully adjust the focus using the **fine focus knob**. Move to the **high-power (40X) objective** and use the **fine focus knob** to adjust the focus.
9. When you are done,
    a. take the slide off the stage and clean it as directed,
    b. turn off the light,
    c. rotate the objectives to the low-power objective, and
    d. cover the scope with the dustcover and store it as directed.

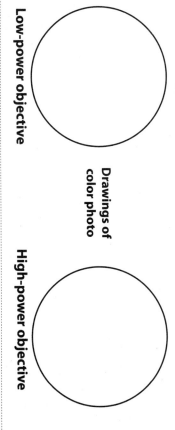

Drawings of color photo

Low-power objective

High-power objective

Microscope Images

Part 1: Draw the letter e under low power.

1. Set the objective lens to 4X.
2. Place the slide of the word *seed* on the stage of the microscope *so you can read it*.
3. Center the image in the field of view on one *e* and draw *exactly* what you see.

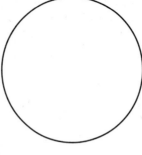

Field of view

Part 2: Move the slide.

4. Move the slide away from you. What direction did the image move? _____
5. Move the slide to your right. What direction did the image move? _____

Part 3: Redraw the letter e after discussion.

6. *After* the class discussion, redraw the letter *e* here.

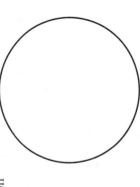

Field of view

Part 4: Answer these questions in your science notebook.

7. Is the image seen through the microscope oriented the same way as the object on the stage of the microscope? Explain.
8. If you want to move the image to the right, which way should you move the slide?
9. If you want to move the image up, which way should you move the slide?

Microscope Images

Part 1: Draw the letter e under low power.

1. Set the objective lens to 4X.
2. Place the slide of the word *seed* on the stage of the microscope *so you can read it*.
3. Center the image in the field of view on one *e* and draw *exactly* what you see.

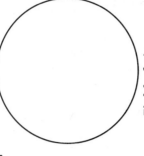

Field of view

Part 2: Move the slide.

4. Move the slide away from you. What direction did the image move? _____
5. Move the slide to your right. What direction did the image move? _____

Part 3: Redraw the letter e after discussion.

6. *After* the class discussion, redraw the letter *e* here.

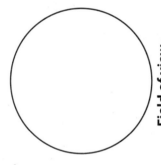

Field of view

Part 4: Answer these questions in your science notebook.

7. Is the image seen through the microscope oriented the same way as the object on the stage of the microscope? Explain.
8. If you want to move the image to the right, which way should you move the slide?
9. If you want to move the image up, which way should you move the slide?

Field of View and Magnification

Part 1: Use the 4X objective.

1. At low power, what is the width of the field of view? _____
2. What is the total magnification with this objective lens? _____

Part 2: Use the 10X objective.

3. Place the slide on the stage and place the clear millimeter ruler on top, using the frame of reference your class decided upon.
4. Draw exactly what you see.

   At medium power, what is the width of the field of view? _____

   What is the total magnification? _____

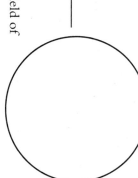

Part 3: Use the 40X objective.

5. Change to high power.
6. Draw exactly what you see.

   At high power, what is the width of the field of view? _____

   What is the total magnification? _____

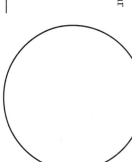

Part 4: Mark the scale.

7. In part 2, for the 10X objective, mark the scale in millimeters on the line under the field of view.
8. In part 3, for the 40X objective, mark the scale in *tenths of millimeters* on the line under the field of view. (Careful!)

Part 5: How big is the letter e?

9. Refer to your notebook and estimate the width of the letter *e* you drew earlier under low power.

   - How wide would the *e* appear under medium power? _____
   - How wide would the *e* appear under high power? _____

Field of View and Magnification

Part 1: Use the 4X objective.

1. At low power, what is the width of the field of view? _____
2. What is the total magnification with this objective lens? _____

Part 2: Use the 10X objective.

3. Place the slide on the stage and place the clear millimeter ruler on top, using the frame of reference your class decided upon.
4. Draw exactly what you see.

   At medium power, what is the width of the field of view? _____

   What is the total magnification? _____

Part 3: Use the 40X objective.

5. Change to high power.
6. Draw exactly what you see.

   At high power, what is the width of the field of view? _____

   What is the total magnification? _____

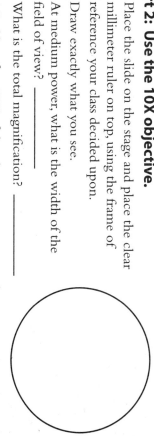

Part 4: Mark the scale.

7. In part 2, for the 10X objective, mark the scale in millimeters on the line under the field of view.
8. In part 3, for the 40X objective, mark the scale in *tenths of millimeters* on the line under the field of view. (Careful!)

Part 5: How big is the letter e?

9. Refer to your notebook and estimate the width of the letter *e* you drew earlier under low power.

   - How wide would the *e* appear under medium power? _____
   - How wide would the *e* appear under high power? _____

Estimating Size

A student determined the diameter of the field of view (FOV) of his microscope at each magnification. He drew an amoeba at each magnification and then drew a follicle mite at each magnification. Estimate the diameter of the FOV and how long each organism is at each magnification.

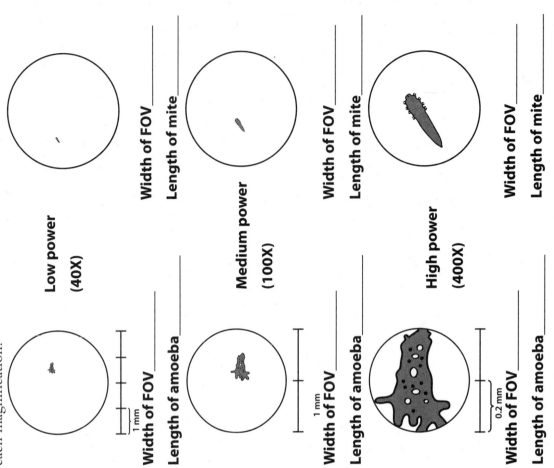

Response Sheet—Investigation 2

A student told her friend,

We looked at things through the microscope today. I saw something called a mite. It got bigger and bigger and bigger as I increased the magnification.

The student's friend looked confused and said,

I am not sure what you mean by "bigger and bigger."

How should the first student correctly explain herself?

Brine Shrimp

WARNING — This set contains chemicals that may be harmful if misused. Read cautions on individual containers carefully. Not to be used by children except under adult supervision.

Part 1: Observe brine shrimp in the vial.

1. How do brine shrimp respond to light? See your teacher's demonstration or shine a flashlight through the vial.

2. Compare the size of the brine shrimp now to the size of the brine shrimp when they first hatched. How are they different?

Part 2: Observe brine shrimp under the microscope.

3. Use a dropper to take up a few brine shrimp. Put one drop on the surface of a slide. If no brine shrimp are on the slide, wipe the slide dry and put on another drop.

4. Use a piece of blotter paper to soak up part of the water.

5. Do not put a coverslip on the slide.

6. Observe and draw an illustration of the brine shrimp at **100X**.

7. Estimate the size of the brine shrimp. _____

Medium power (100X)

Part 3: Add yeast to the brine shrimp.

8. Carefully add one tiny drop of Congo red–dyed yeast to the slide.

9. Observe the yeast and the brine shrimp. Describe what you see.

Looking at Elodea

Part 1: Observe elodea leaf layers.

1. Place a small elodea leaf on a slide, top side up, bottom side against the slide. Prepare a wet mount using pond water and a coverslip.
2. Focus the microscope at 40X and then increase to 100X.
3. Increase the magnification to 400X. Using the fine focus knob, carefully focus up and down through the different layers of the leaf. How many layers can you see? _____
4. Describe what you observe.

Part 2: Observe elodea details and cell size.

5. Look carefully for movement inside the leaf. Describe what you observe.
6. Draw a few representative *large* brick-like structures to scale in the circle. Do not fill in the entire field of view. Use color and include detail.

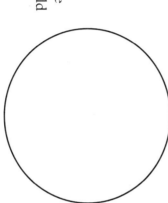

High power (400X)

7. How many of the *large* green "bricks" fit lengthwise across the field of view?
8. Estimate the size of one of the "bricks." _____

Part 3: Label the drawing.

9. Label the cell wall, chloroplasts, and cytoplasm.

Plant Cell Structures and Functions

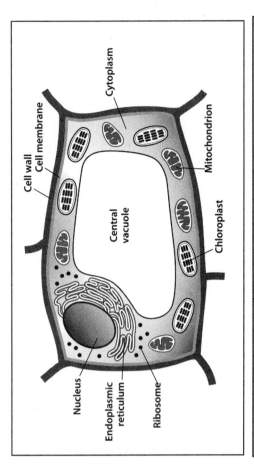

Cell structure	Function
Cell membrane	
Cell wall	
Central vacuole	
Chloroplast	
Cytoplasm	
Endoplasmic reticulum	
Mitochondrion	
Nucleus	
Ribosome	

Plant Cell Structures and Functions

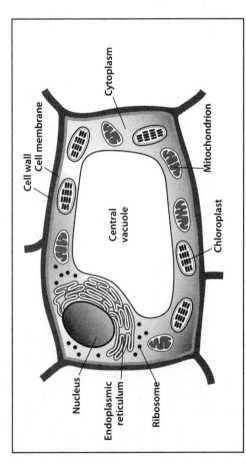

Cell structure	Function
Cell membrane	
Cell wall	
Central vacuole	
Chloroplast	
Cytoplasm	
Endoplasmic reticulum	
Mitochondrion	
Nucleus	
Ribosome	

Paramecia

WARNING — This set contains chemicals that may be harmful if misused. Read cautions on individual containers carefully. Not to be used by children except under adult supervision.

Part 1: Observe movement and behavior.

1. Put one small drop of paramecium culture on the center of your slide. Do NOT put on a coverslip.
2. Focus the microscope at 40X to make sure you have paramecia on your slide. Increase the magnification to 100X.
3. Describe the movement and behavior of the paramecia.

Part 2: Observe paramecium up close.

4. Remove the slide from the stage and add one drop of methyl cellulose. Put on a coverslip. If necessary, blot up extra liquid.
5. Find one paramecium that is still moving, focus under low power, and increase to medium and then to high power. Focus using the fine focus knob. Describe the paramecium and draw it in the circle below.
6. Estimate the length of the paramecium. _____

Part 3: Label the drawing.

7. Label the cell membrane, cytoplasm, cilia, and any other structures you observe.
8. What is the purpose of the cell membrane?

High power (400X)

Protist Cell Structures and Functions

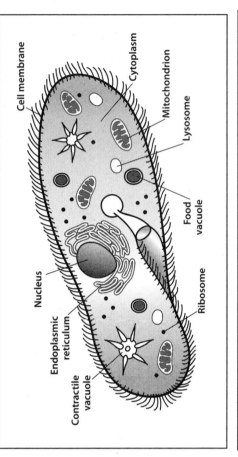

Cell structure	Function
	Boundary that controls what enters and leaves the cell
	A membrane that stores water and expels excess water
	Internal fluid that contains the cell structures
	A membranous structure that assembles proteins and parts of the cell membrane
	A membrane that stores food and merges with a lysosome to digest food
	An organelle that digests cellular waste and merges with a food vacuole to digest food
	An organelle that converts the energy in food into usable energy for the cell
	An organelle that contains the cell's genetic material (DNA), which determines the nature of cell structures and substances
	A structure that makes proteins, either free or bound to the endoplasmic reticulum

Protist Cell Structures and Functions

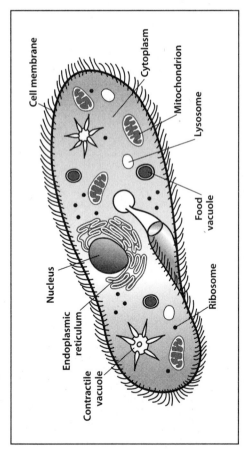

Cell structure	Function
	Boundary that controls what enters and leaves the cell
	A membrane that stores water and expels excess water
	Internal fluid that contains the cell structures
	A membranous structure that assembles proteins and parts of the cell membrane
	A membrane that stores food and merges with a lysosome to digest food
	An organelle that digests cellular waste and merges with a food vacuole to digest food
	An organelle that converts the energy in food into usable energy for the cell
	An organelle that contains the cell's genetic material (DNA), which determines the nature of cell structures and substances
	A structure that makes proteins, either free or bound to the endoplasmic reticulum

Response Sheet—Investigation 3

Two students were having a discussion. One said,

All cells are living things. Every cell in an elodea plant is an organism, just like the one-celled paramecium we looked at.

The second student said,

Well, you're partly right. I agree that all cells are living things, but an elodea cell is not an organism.

Evaluate what each student said. Explain your thinking.

First student:

Second student:

WARNING — This set contains chemicals that may be harmful if misused. Read cautions on individual containers carefully. Not to be used by children except under adult supervision.

Minihabitat Safari

Is there anything living in the minihabitat?

1. Prepare a wet mount from one region of your minihabitat. Look for life at 40X.
2. If necessary, add one drop of methyl cellulose. Put on a coverslip and blot away any extra liquid. Increase the magnification to 100X and then 400X as needed.
3. Draw to scale any organisms you observe. Use the next page in your science notebook to describe their behavior and to add more organisms.
4. Use "Microorganism Guide" in *Science Resources* to help identify any organisms you find.

Organism _____
Estimated size _____

Organism _____
Estimated size _____

Organism _____
Estimated size _____

Organism _____
Estimated size _____

Human Cheek Tissue

WARNING — This set contains chemicals that may be harmful if misused. Read cautions on individual containers carefully. Not to be used by children except under adult supervision.

Part 1: Prepare a cheek-tissue sample.

1. Gently rub the inside of your cheek with a cotton swab.
2. Roll the rubbing onto the center of a slide. Add one drop of methylene blue and let set for 1 minute.
3. Hold the slide over a waste container and rinse it with a few drops of water. Add a drop of water if necessary, place a coverslip on top, and blot any extra water from the edges.
4. View the slide, starting at 40X. Use the search image your teacher provides to help you focus on the stained cheek tissue. Increase magnification to 400X.

Part 2: Record observations.

5. Describe what you see at 400X and draw it in the circle.

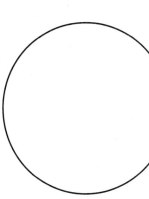

High power (400X)

Part 3: Questions

6. Estimate the diameter of one cell.
7. What is the inside of your cheek made of?
8. What do you think other parts of your body are made of?
9. Label the cell membrane and nucleus in one of the cheek cells.

Clean up as directed.

Animal Cell Structures and Functions

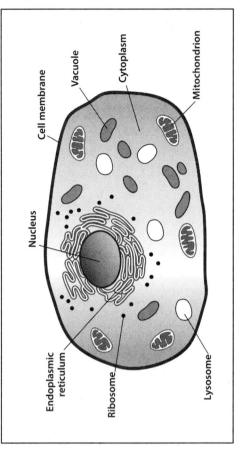

Cell structure	Function
	Boundary that controls what enters and leaves the cell
	Internal fluid that contains the cell structures
	A membranous structure that assembles proteins and parts of the cell membrane
	An organelle that digests cellular waste
	An organelle that converts the energy in food into usable energy for the cell
	An organelle that contains the cell's genetic material (DNA), which determines the nature of cell structures and substances
	A structure that makes proteins, either free or bound to the surface of the endoplasmic reticulum
	A membrane that stores water and other materials

Animal Cell Structures and Functions

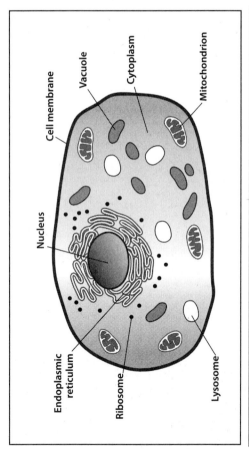

Cell structure	Function
	Boundary that controls what enters and leaves the cell
	Internal fluid that contains the cell structures
	A membranous structure that assembles proteins and parts of the cell membrane
	An organelle that digests cellular waste
	An organelle that converts the energy in food into usable energy for the cell
	An organelle that contains the cell's genetic material (DNA), which determines the nature of cell structures and substances
	A structure that makes proteins, either free or bound to the surface of the endoplasmic reticulum
	A membrane that stores water and other materials

Cells

"Why should I care about cells?" That's a reasonable question.

The Importance of Cells

You can't see them, and you might not have heard much about them before, so why do they matter?

Well, for one reason, life on Earth exists as cells.

"Wait," you say, "I am certainly not a cell." And you are 100 percent correct. But you are made of cells, such as nerve cells, liver cells, lung cells, blood cells, muscle cells, and many more.

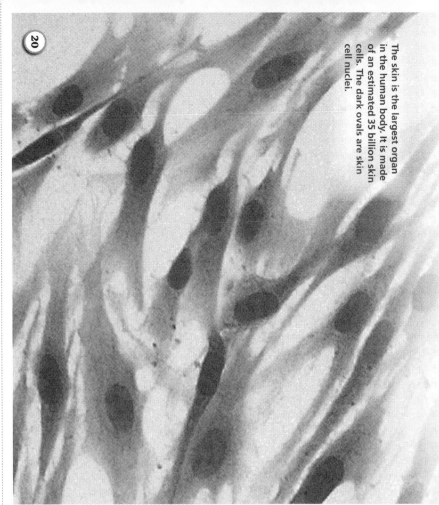

The skin is the largest organ in the human body. It is made of an estimated 35 billion skin cells. The dark ovals are skin cell nuclei.

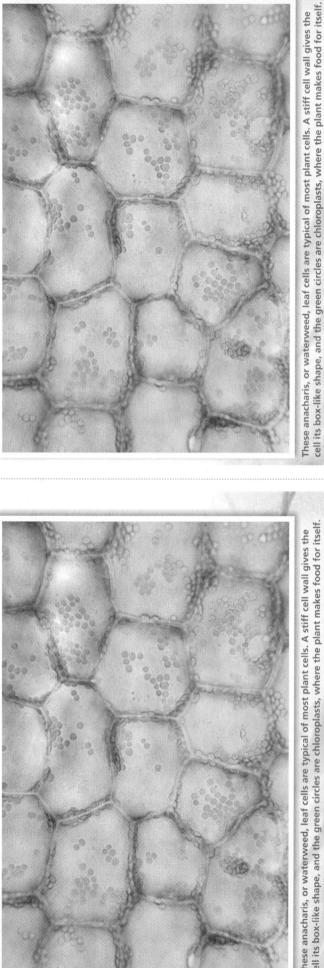

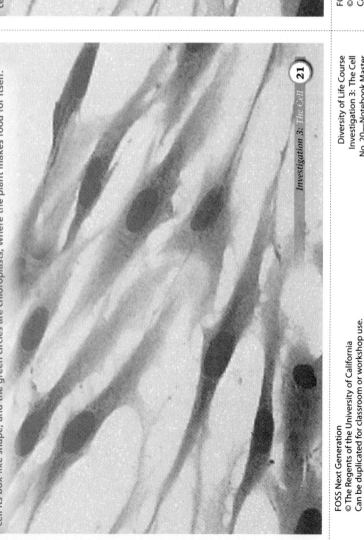

These anacharis, or waterweed, leaf cells are typical of most plant cells. A stiff cell wall gives the cell its box-like shape, and the green circles are chloroplasts, where the plant makes food for itself.

Not only that, every living thing is made of cells. Some organisms, including **insects**, trees, worms, mushrooms, and the neighborhood cat, are **multicellular organisms**. That means they are made of millions, even trillions, of cells. In fact, the adult human body is made up of anywhere from 60 to 90 trillion cells.

Most organisms, however, are single-celled or unicellular. They consist of one cell. Organisms such as **bacteria, archaea,** paramecia, and other protists are all made of one cell. That's amazing! How could a single microscopic cell be a living thing? It brings up the question "Are cells living?" What do you think?

All cells exhibit *all* the characteristics of life, even if they are part of a larger organism. Each cell carries out all the work necessary to sustain life. Before we figure out how they do this, let's take a look at how cells were discovered.

Did You Know?

If you lined up all the cells in a human body end to end, you could circle Earth more than four times!

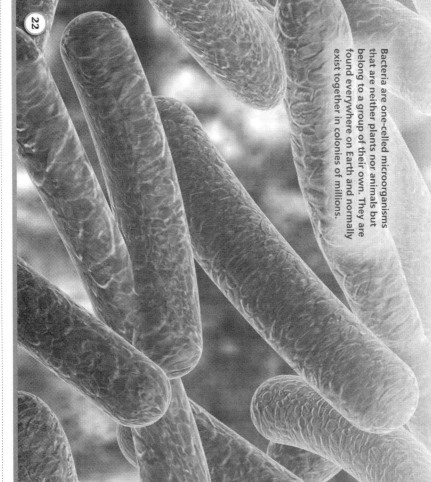

Bacteria are one-celled microorganisms that are neither plants nor animals but belong to a group of their own. They are found everywhere on Earth and normally exist together in colonies of millions.

The Discovery of Cells

Until the 1600s, no one had any idea that all life is made of cells. There was no way to see cells. The technology did not exist until eyeglass makers in Europe developed lenses that magnified what they were looking at. This discovery led to the invention of the compound microscope.

The invention of microscopes led to the discovery of cells. In 1665, Robert Hooke (1635–1703) looked at a sample of cork under the microscope. What he saw in the **field of view** reminded him of rows of tiny rooms in a monastery, or cells. He drew the cells to **scale** in order to identify their structures and start to analyze their functions.

Around the same time, Antoni van Leeuwenhoek looked at samples of clear pond water with his microscope. He was amazed to find tiny things swimming around. His definition of life was not as sophisticated as the one we use now, but he concluded correctly that those swimming things were living organisms. He soon found them everywhere, even in his own mouth.

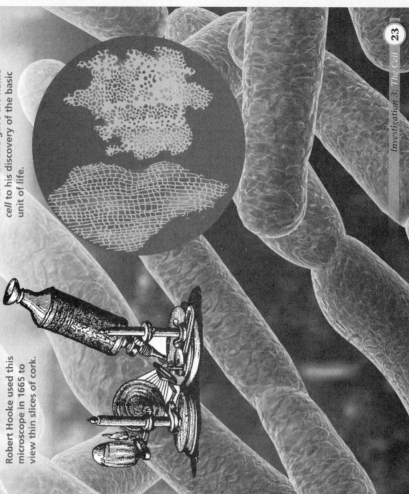

Robert Hooke used this microscope in 1665 to view thin slices of cork.

Hooke sketched the "little rooms" he observed and gave the name *cell* to his discovery of the basic unit of life.

The Cell Theory

Before long, scientific observations led to the conclusion that cells are the basic units of life. In the 1830s, biologists such as Matthias Schleiden (1804–1881), Theodor Schwann (1810–1882), and Rudolf Virchow (1821–1902) concluded that all plants and animals are made of cells. Soon it was confirmed that new cells come only from the division of existing cells. These conclusions led scientists to summarize their findings in what is called the cell theory. The cell theory states,

1. All living things are made up of one or more cells.
2. Cells are the basic units of structure and function in living things.
3. All living cells come from existing cells.

Take Note

Write the three statements of the cell theory in your science notebook.

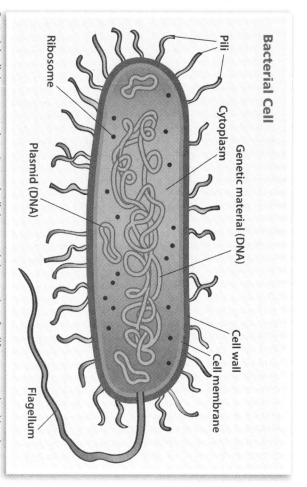

Bacterial Cell

Bacterial cells have no nucleus. Instead, all the genetic instructions for life are contained in a long strand of DNA. The only other structures in a bacterial cell are ribosomes, where protein is made.

Different Types of Cells

Cells come in many shapes and sizes. Some are as small as 0.2 micrometers (μm) across, and some as large as 10 centimeters (cm) across. Some are individual living organisms. Others are the smallest living parts of much larger multicellular organisms.

All cells are separated from their environment by a cell membrane. This porous flexible structure allows some things to pass in and keeps other things out. This boundary keeps the inside fluid part of the cell, the **cytoplasm**, contained.

Cells such as bacteria and archaea are called **prokaryotes**. There is very little apparent organization of the materials inside their cell membrane. In fact, their **genetic material**, in the form of deoxyribonucleic acid (DNA) or ribonucleic acid (RNA), simply floats in the cytoplasm. This is the main characteristic that defines prokaryotes. Prokaryotic cells carry out the business of life with very few specific cell structures.

Organisms with more complex cells (including all multicellular organisms) are called **eukaryotes**. In eukaryotic cells, the cytoplasm contains cell structures called organelles, meaning "little organs." Just as the human body is made up of organs and **organ systems**, which take care of life's functions, eukaryotic cells are made up of organelles, each of which has a job to do.

Protist Cell

Like bacteria, protists are single-celled microorganisms. Unlike bacteria, protist cells have a nucleus and many specialized organelles for carrying out specific life functions.

Animal Cell

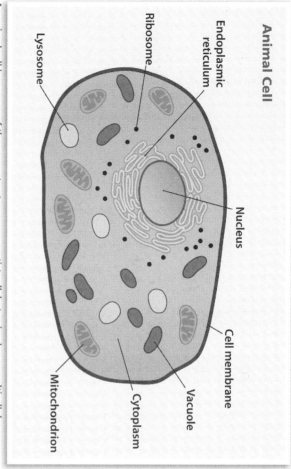

An animal cell has many of the same structures as a protist cell, but animals are multicellular, or made up of many cells. All the functions of life are carried out inside every cell.

How Do Cells Carry Out Life's Functions?

Cells exchange gases. Gases such as carbon dioxide and oxygen move through the cell membrane.

Cells need water. In fact, all cells are found in an **aquatic** environment, a water-based fluid. Just like gases, water flows back and forth across the cell membrane. Water is necessary for all chemical processes that happen in cells.

Cells need food. Plant cells make their own food in organelles called **chloroplasts**. Single-celled paramecia eat other **microorganisms**. Humans eat vegetables, fruits, and meat. Some bacteria consume sulfur. In eukaryotic cells, an organelle called the **mitochondrion** processes food to obtain usable energy.

Cells eliminate waste. The **lysosome** is the animal cell's waste-disposal organelle. Plant cells digest unwanted material in the central vacuole before discarding it through the cell membrane as waste.

Cells reproduce. The **nucleus** in eukaryotic cells contains the genetic information that drives cell division. And even though prokaryotes don't have a nucleus, they still have genetic material.

Cells grow. Several cell structures are involved in making proteins, which are used to build new cell structures. **Ribosomes** take

Plant Cell

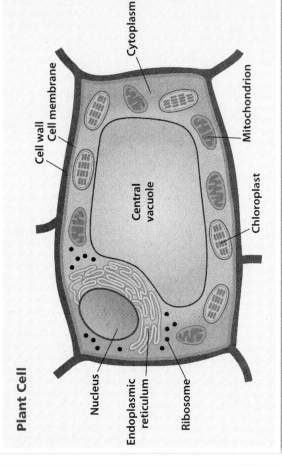

A plant cell has important structures that an animal cell lacks: a cell wall, which provides support and structure, and chloroplasts, which produce food by the process of photosynthesis.

part in protein synthesis. In eukaryotic cells, the ribosomes may be attached to a structure called the rough **endoplasmic reticulum**.

Cells respond to their environment. Paramecia swim around in search of food. They move away from cold or hot areas. Most cells respond to chemicals in their environment.

Cells need a suitable environment. If the environment around cells becomes toxic, they die. Some cells form thick protective walls called cysts to protect themselves when an environment becomes stressful. When the environment improves, the cysts break open, and the cells resume their normal lives.

Why Should I Care?

Perhaps it is obvious by now. You, your friends, your family members, and every other living thing on Earth are composed of one or more cells. If cells did not exist, life would not exist. You would not exist!

Think Questions

1. **How did changes in technology lead to the discovery of cells?**
2. **Describe how eukaryotic cells and prokaryotic cells are similar and how they are different.**
3. **Why are cells considered one of the characteristics of life?**

Observing Bacteria

Date	Sample taken from	Description of agar plate (include number of colonies)	Drawing of agar plate (include color)
		Observations 1. 2. 3. 4.	1 | 2 --- 3 | 4
		Observations 1. 2. 3. 4.	1 | 2 --- 3 | 4
		Observations 1. 2. 3. 4.	1 | 2 --- 3 | 4
		Observations 1. 2. 3. 4.	1 | 2 --- 3 | 4

Observing Bacteria

Date	Sample taken from	Description of agar plate (include number of colonies)	Drawing of agar plate (include color)
		Observations 1. 2. 3. 4.	1 | 2 --- 3 | 4
		Observations 1. 2. 3. 4.	1 | 2 --- 3 | 4
		Observations 1. 2. 3. 4.	1 | 2 --- 3 | 4
		Observations 1. 2. 3. 4.	1 | 2 --- 3 | 4

Observing Fungi

Date	Description of bread (include number of colonies)	Drawing of bread
	Sample taken from	
	Observations	
	Observations	
	Observations	

Observing Fungi

Date	Description of bread (include number of colonies)	Drawing of bread
	Sample taken from	
	Observations	
	Observations	
	Observations	

Response Sheet—Investigation 4

A student wrote the following response to the question, What are elodea plants made of?

Elodea plants are made of cells, cell walls, cytoplasm, and chloroplasts.

His friend told him that he forgot to include the levels of complexity. Improve on the first student's response, keeping in mind his friend's suggestion.

Bacterial Cell Structures and Functions

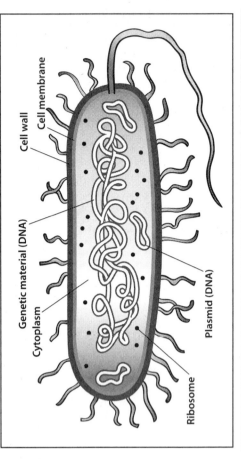

Cell structure	Function
	Boundary that controls what enters and leaves the cell
	A rigid layer that supports the cell and provides shape
	Internal fluid that contains the cell structures
	A molecule that determines the nature of cell structures and substances
	Small piece of genetic material that is independent of other DNA in the cell and that can be passed to other bacteria
	A structure that makes proteins, in the cytoplasm

Bacterial Cell Structures and Functions

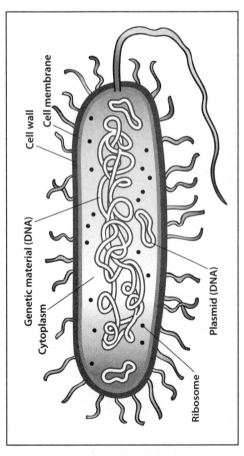

Cell structure	Function
	Boundary that controls what enters and leaves the cell
	A rigid layer that supports the cell and provides shape
	Internal fluid that contains the cell structures
	A molecule that determines the nature of cell structures and substances
	Small piece of genetic material that is independent of other DNA in the cell and that can be passed to other bacteria
	A structure that makes proteins, in the cytoplasm

Fungal Cell Structures and Functions

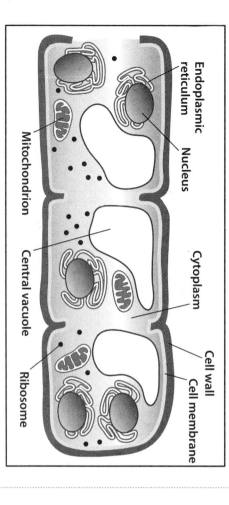

Cell structure	Function
	Boundary that controls what enters and leaves the cell
	A rigid layer that supports the cell and provides shape
	Internal fluid that contains the cell structures
	A membranous structure that assembles proteins and parts of the cell membrane
	Converts the energy in food into usable energy for the cell
	Contains the cell's genetic material (DNA), which determines the nature of cell structures and substances
	A structure that makes proteins, free or bound to the surface of the endoplasmic reticulum
	A membrane that stores water and other substances, and provides structure and support for the cell

---

Fungal Cell Structures and Functions

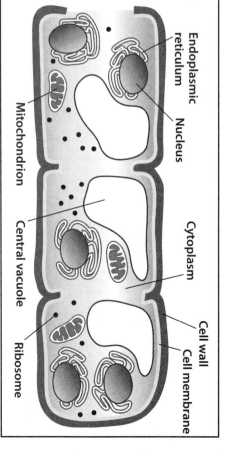

Cell structure	Function
	Boundary that controls what enters and leaves the cell
	A rigid layer that supports the cell and provides shape
	Internal fluid that contains the cell structures
	A membranous structure that assembles proteins and parts of the cell membrane
	Converts the energy in food into usable energy for the cell
	Contains the cell's genetic material (DNA), which determines the nature of cell structures and substances
	A structure that makes proteins, free or bound to the surface of the endoplasmic reticulum
	A membrane that stores water and other substances, and provides structure and support for the cell

Archaea

Archaea have the same cell structures as bacteria.

1. Do archaea have a nucleus? _____

2. Do archaea have DNA and plasmids? _____

3. Do archaea have cytoplasm? _____

4. Do archaea have a cell wall and cell membrane? _____

5. Do archaea have ribosomes? _____

6. What is unique about archaea? During the class discussion, record your notes below.

Classification History Notes

Answer these questions in your notebook.

1. Aristotle thought that all life had _____. He thought the lowest form of life was _____ and the highest form of life was _____ because they can reason. Linnaeus identified two kingdoms: _____ and _____.

2. The idea of two kingdoms remained in place until _____. The _____ enabled scientists to identify single-celled organisms.

3. It wasn't until the 1950s and 1960s that scientists classified life into five kingdoms because some kinds of life did not match plants and animals. The three new kingdoms were _____, _____, and _____.

4. A further classification of life depended on whether there is a nucleus in the cell. What are the two kinds of cells called? Give examples of each kind.

5. Another change was proposed in the 1970s, when the technology was sophisticated enough to analyze molecular structures inside cells. The current classification of life is now based on three domains: Bacteria, Archaea, and Eukaryota (cells with a nucleus and organelles). On the opposite page in your notebook, draw how the three domains are related and label each one with the organisms that belong in them. This is our current understanding of how life is organized.

Celery Investigation A

	Day 0 (set-up)	Day 1 (final)	Change
Water in control vial (volume)			
Water in celery vial (volume)			
Celery mass			

Part 1

1. Take the celery stalk out of the vial. Measure the amount of water in the celery vial, using a graduated cylinder. Record.

2. Record the amount of water in the class control (evaporation) vial.

3. Calculate the changes in volume of water in the vials. Record.

4. How much water was lost to evaporation? _____

5. How much water was lost in the celery vial? _____

6. Do the amounts match? _____ Why or why not?

Part 2

7. *Predict* the current mass of the celery stalk. (Remember that for water, 1 mL = 1 g.) _____

8. Determine the actual mass of the celery. Record.

Celery Investigation B

Part 2 (continued)

9. Does your prediction match the actual mass of the celery? _____ Record any ideas you have about your results.

10. Calculate the *change* in mass of the celery. Record in the data table.

Part 3

11. How much of the water from the celery vial ended up in the celery? _____ How do you know?

12. What do you think happened to the rest of the water that was lost from the celery vial?

13. Determine the amount of water unaccounted for in your vial.

water lost from celery vial — water that evaporated — any *increase* in mass of the celery = water unaccounted for

Part 4

14. In your notebook, describe any patterns you notice in the class celery and class data.

Leaf Observations

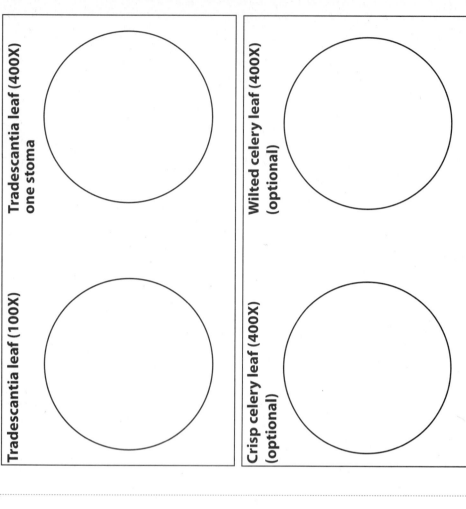

Tradescantia leaf (100X)

Tradescantia leaf (400X) one stoma

Crisp celery leaf (400X) (optional)

Wilted celery leaf (400X) (optional)

1. Label the guard cells and one stoma in the high-power drawing.
2. Describe the structure of a stoma.
3. Explain how stomata work.

Response Sheet—Investigation 5

A student noticed a plant outside that had really wilted leaves. He remarked to a friend,

Those leaves must be losing a lot of water to become so wilted. I bet that the stomata are totally open right now.

Do you agree or disagree? What would you add to the conversation?

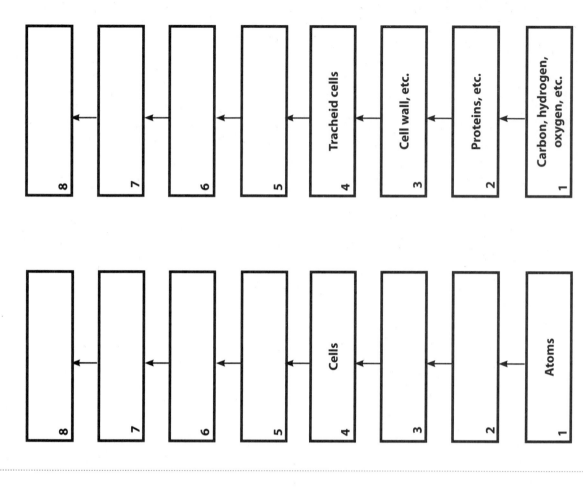
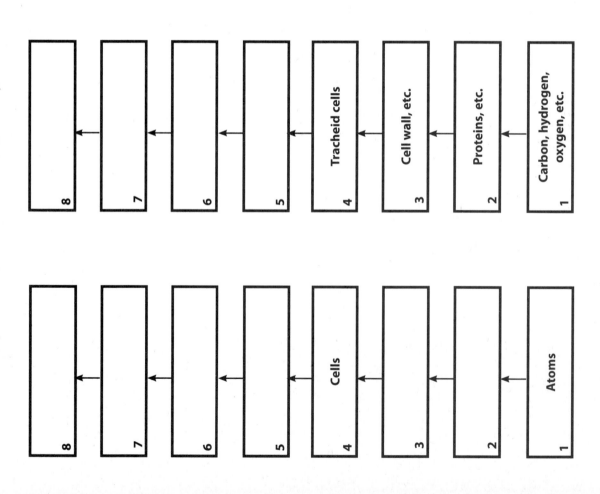

Water, Light, and Energy

It was your chore to water the houseplants, but you forgot! When you finally remembered, it was too late. The plants looked brown and shriveled.

Needs of Plants

Plants need water in order to live. "Needs water" is on the list of life requirements for *all* life.

Have you ever seen houseplants wilt, lose their healthy green color, and die if they don't get enough light? "Needs light" isn't on our list. Is light one of the requirements for all life? Should we add it to the list, or can we figure out how plants' need for light fits into the list we already have?

> **Take Note**
> Record in your notebook your ideas about whether "needs light" should be added to the list of requirements for life.

Dry soil is by far the most common reason why plants wilt. A good dousing with water will usually revive them.

All Plant Cells Need Water

The answer to our question about light starts with water. Plants use water to transport minerals to all their cells. Any substances in a biochemical reaction must be dissolved in water. Plants also use water to cool off in the heat of the day, to give them shape, and to grow. Plant cells are filled with cytoplasm, which is mostly water.

Plants get the water they need from the soil. **Root hairs** take up water and pass it into cellular tubes, which make up **xylem** tissue, part of the plant's **vascular system**. As we saw in the celery investigation, xylem tubes carry water up through the stem to the smaller **veins** in the leaves. In this way, all the cells in a plant get water and minerals from the soil. Xylem tubes are made of the cell walls of dead xylem cells. The cells are connected end to end in the stems, like long straws. The tubes form a pipe system that can be extremely long and complex, especially in huge trees like the giant sequoia.

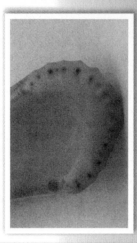

Xylem tubes in celery carry water and minerals from the plant's roots up through the stalk.

Veins are vascular bundles that form transport networks though the plant's leaves.

Water loss out of leaves by evaporation is called transpiration. This process pulls water upward through the plant.

Transpiration

What happens when water finally reaches the leaves? What did you observe in the bag around a leafy twig? Water! Where did that water come from? It came from the plant. It left the plant leaves as water vapor and entered the atmosphere. This process is called transpiration. Water vapor (lots of it) exits the leaves through small pores called stomata (stoma = mouth). Guard cells open and close the stomata. They control the movement of gases, including water vapor, into and out of the leaf. But not all water exits the plant.

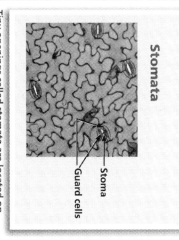

Stomata

Tiny openings called stomata are located on the underside of leaves. Guard cells surround and regulate gas and moisture exchange through these "doorways."

Water is one of the ingredients a plant needs to produce food. Later, when the plant uses its food for energy, it produces water as a by-product!

Photosynthesis

You have probably heard that plants and **algae** (and some bacteria) produce their own food. How do they do it and what do water and light have do to with it? The process of photosynthesis is the answer. Thinking about photosynthesis will help us answer the question about light. Water is involved in the photosynthesis process in at least two ways. Water dissolves substances to make them available for chemical reactions. One of those substances is carbon dioxide gas. The stomata open for gas exchange, allowing carbon dioxide from the atmosphere to enter.

The carbon dioxide dissolves in water in the spaces surrounding the cells. Dissolved carbon dioxide enters the nearby cells, where it becomes one of the building blocks of sugar. So water makes carbon dioxide available.

Water is also important because it is the other building block of sugar. Water combines chemically with carbon dioxide to produce sugar. Sugar provides food energy for the plant. And it provides food for any other living thing that eats the plant.

Only one thing is missing. What do you think it is? The other thing plants need to make their own food is light.

Chloroplasts are even found in guard cells, although food production is not the guard cells' primary function.

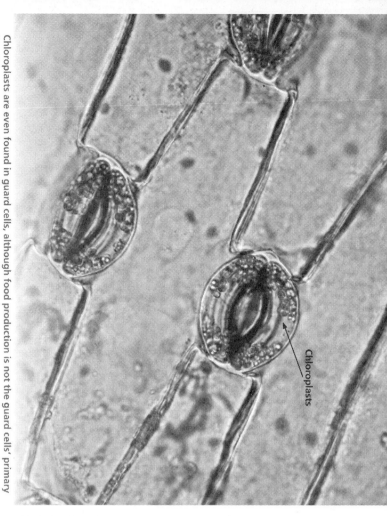
Chloroplasts

$6CO_2 + 6H_2O + \text{light energy} \longrightarrow C_6H_{12}O_6 + 6O_2$
Carbon dioxide + water + light energy *produces* **sugar + oxygen**

The overall simplified chemical reaction can be expressed this way.

This reaction occurs only in the chloroplasts, the green organelles that you observed in **elodea** leaf cells. You cannot just dissolve some carbon dioxide, throw it together with water, hit it with light energy, and expect to make sugars. The reaction happens only in the chloroplasts. A green chemical pigment called **chlorophyll** allows the plant to capture and convert light energy into chemical energy of the bonds in sugar.

In order to make their own food, plants need water, carbon dioxide, light energy in the form of sunlight, and chlorophyll. The process is called photosynthesis, which makes sense because *photo* means light, and *synthesis* means putting together.

Aerobic Cellular Respiration

Now we know how plant cells that have chloroplasts make food. But the cells in plant roots do not have chloroplasts. How do they get the food they need? The food made by the cells with chlorophyll must somehow get to the root cells. Next to the xylem is another part of the plant's vascular system called **phloem** tubes. These tubes carry sugar from the leaves to all the other cells of the plant. All the cells of a plant get food delivered via the phloem. But the energy stored in sugar is not in a form plants can use to grow, repair damaged tissue, or make new structures. In order to change sugar to a form that cells *can* use, plants need oxygen. You saw that during photosynthesis, plants use carbon dioxide and give off oxygen. But plants need oxygen, too.

Like most living cells, plant cells use oxygen to transform the energy in sugars into a usable form of energy. Oxygen and sugar (a molecule called glucose), combine to release energy, and carbon dioxide and water are produced as waste. This reaction happens in every plant cell, every animal cell, and almost all other living cells.

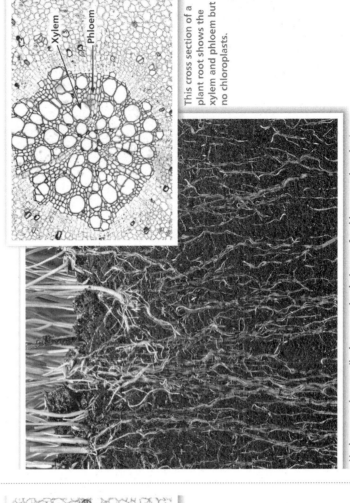

This cross section of a plant root shows the xylem and phloem but no chloroplasts.

Underground root cells do not make their own food because they don't have chlorophyll.

$C_6H_{12}O_6 + 6O_2 \rightarrow 6CO_2 + 6H_2O + energy$

Glucose + oxygen *produces* **carbon dioxide + water + energy**

The chemical reaction can be expressed this way.

In eukaryotic cells, this process happens in the mitochondria. It is called **aerobic cellular respiration** (*aerobic* means it uses oxygen). Notice anything interesting? Compare the equation for photosynthesis and the equation for cellular respiration. The sugar molecules are on opposite sides of the equations, and the light energy became usable energy for the plant. Almost all organisms rely on aerobic cellular respiration to convert glucose into usable energy.

But only photosynthetic organisms can capture the Sun's energy to create sugars. How do other organisms get sugars? All other organisms must eat photosynthetic organisms, such as plants, or eat organisms that did so.

> **Take Note**
> How are photosynthesis and aerobic cellular respiration alike? How are they different?

Without the Sun, we'd have a hungry planet. Sunlight provides the energy needed for photosynthesis. In this chemical process, plants make food for themselves, for the animals that eat them, and for the animals that eat those animals.

Summary

Plants need water. They pull water up from soil, using transpiration. Plants transport water to all cells, using xylem tubes in their vascular system. Water vapor exits leaves through stomata, which guard cells open and close. Plants make their own food from water and carbon dioxide. This process, called photosynthesis, uses light and chlorophyll. The food they make is in the form of sugar (glucose), which stores energy. Plants transport sugar to all cells, using phloem in the vascular system. Plants, like most life-forms, use aerobic cellular respiration to change sugar into usable energy to perform all of life's functions.

We do not need to add "needs light" to our list of life requirements. The light that plants need is a part of "needs food." So water those plants, and make sure that they get the light they need!

Think Questions

1. Explain why water is necessary for plants to make food.
2. How do all the cells in a plant get the water they need? Explain.
3. Do all plant and animal cells photosynthesize? Explain.
4. Do all plant and animal cells use aerobic cellular respiration? Explain.
5. Is light a requirement for life? Explain.

Seed Dissection

Dry seed dissection: Draw and label what you observe.

Outside of seed	Inside of seed

Soaked seed dissection: Draw and label what you observe.

Outside of seed	Inside of seed

Answer these questions in your notebook.

1. How is a seed protected during dormancy?
2. If a seed did not have cotyledons, what would happen?
3. Why do you think the ability to produce seeds is an important adaptation for flowering plants?

Germination and Growth in Different Salinities

WARNING — This set contains chemicals that may be harmful if misused. Read cautions on individual containers carefully. Not to be used by children except under adult supervision.

The kind of seed we are investigating: _____

Number of seeds in each dish: _____

1. Record the number of seeds with roots and the number of seeds with shoots in the table below.

    # seeds with roots / # seeds with shoots

	0 spoons salt	1 spoon salt	2 spoons salt	4 spoons salt
Day 2				
Day ___				

2. On the final day, make your observations and comments.

0 spoons salt	1 spoon salt

2 spoons salt	4 spoons salt

WARNING — This set contains chemicals that may be harmful if misused. Read cautions on individual containers carefully. Not to be used by children except under adult supervision.

Comparing Growth

Part 1: Think about the seeds you investigated.

The kind of seed we are investigating: _____

1. In which condition(s) did most of your seeds germinate?

   In which condition(s) did the fewest of your seeds germinate?

2. In which condition(s) do the roots and the shoots of your seeds appear the healthiest? (Compare length of roots and shoots, branching of roots, number of root hairs, greenness.)

3. How does increasing the concentration of salt affect the germination and growth of your seeds?

Part 2: Compare all the seeds at each concentration of salt.

4. Which seeds (oats, wheat, barley, or corn) grew the best at 0 spoons, 1 spoon, 2 spoons, 4 spoons of salt? (Compare number of seeds germinated, healthiest looking.)

	0 spoons salt	1 spoon salt	2 spoons salt	4 spoons salt
Seed type showing most salt tolerance				

5. Which type of food crop is best suited to saline (salty) soil?

6. Answer in your notebook: Is saline soil a suitable environment for germinating and growing food crops? What is your evidence?

Parts of a Flower

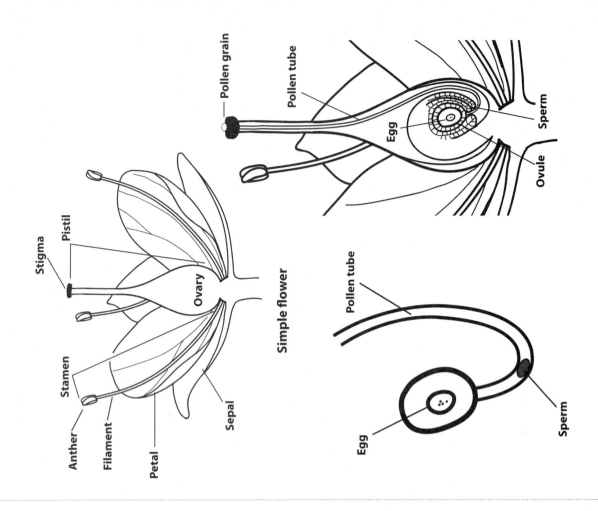

Flower Dissection A

Dissection of a _____ flower

1. Look into the center of the flower. Draw a picture showing how the stamen and the pistil are arranged. Label your drawing.

2. Observe the end of the stamen closely. Make a close-up drawing showing the structure at the end of the stamen. Label your drawing.

3. Gently push your finger into the center of the flower. Look closely at your finger with a hand lens. Describe what you see.

4. If a microscope is handy, put some of the material on a slide and observe it at 100X and 400X. Draw what you see under high power. Label your drawing.

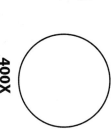

400X

Flower Dissection A

Dissection of a _____ flower

1. Look into the center of the flower. Draw a picture showing how the stamen and the pistil are arranged. Label your drawing.

2. Observe the end of the stamen closely. Make a close-up drawing showing the structure at the end of the stamen. Label your drawing.

3. Gently push your finger into the center of the flower. Look closely at your finger with a hand lens. Describe what you see.

4. If a microscope is handy, put some of the material on a slide and observe it at 100X and 400X. Draw what you see under high power. Label your drawing.

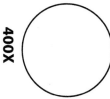

400X

Flower Dissection B

5. Remove the sepals. How many are there? _____ Stick one sepal upside down on the tape near the right end.

6. Remove the petals. How many are there? _____ Stick one petal upside down on the tape next to the sepal.

7. Remove the stamens. How many are there? _____ Put all the stamens on the tape.

8. The remaining part is the pistil, which includes the ovary. Use a hand lens to observe the stigma of the pistil. Draw and label it.

9. Ask your teacher to cut open the ovary. Examine the inside of the ovary with your hand lens. Draw and label what you see.

    Place the pistil with the ovary cut side down on the tape next to the stamens.

10. Slide the card out from under the tape. Center the card on top of the mounted flower parts. Press down firmly to stick the card to the tape. Carefully lift up the ends of the tape and fold them to the back of the card to complete the flower mount. Label all the parts.

Plant-Reproduction Cards

A A pollen grain, usually carried by animal or air, lands on the stigma of another flower.	**F** The sperm cell fertilizes an egg. The egg and sperm fuse to form a single cell with information from the male and female.
B The parent plant forms a food source for the developing embryo.	**G** The pollen grain forms a long tube down the length of the pistil into the ovule.
C A sperm cell travels down the pollen tube.	**H** The seed-containing ovary develops into a fruit.
D Fruit is dropped or consumed by an animal, and the seed is released.	**I** The single cell divides, and each of those cells divides, and so on until the many cells develop into an embryo.
E Pollen grains, which contain the male sperm cells, form on the anthers.	**J** Ovules, which contain the female egg cells, form in the ovary.

Plant-Reproduction Cards

A A pollen grain, usually carried by animal or air, lands on the stigma of another flower.	**F** The sperm cell fertilizes an egg. The egg and sperm fuse to form a single cell with information from the male and female.
B The parent plant forms a food source for the developing embryo.	**G** The pollen grain forms a long tube down the length of the pistil into the ovule.
C A sperm cell travels down the pollen tube.	**H** The seed-containing ovary develops into a fruit.
D Fruit is dropped or consumed by an animal, and the seed is released.	**I** The single cell divides, and each of those cells divides, and so on until the many cells develop into an embryo.
E Pollen grains, which contain the male sperm cells, form on the anthers.	**J** Ovules, which contain the female egg cells, form in the ovary.

Response Sheet—Investigation 6

One of your good friends was absent the day that her class discussed plant reproduction. She is trying to write a paragraph describing flowering-plant reproduction.

All I know is that baby plants come from seeds—I don't know where seeds come from.

What would you tell your friend that would help her understand how flowering plants reproduce?

Pollination Syndrome A

Part 1: Observe your flower.

1. Describe the shape and color of the flower.

2. Describe any scent the flower has.

3. List any other characteristics that you think might attract pollinators.

Part 2: Use the "Flower Information" resource.

Look for an example of a flower that is similar to yours.

4. Where are the anthers and the stigma located in relationship to each other?

5. Where would a pollinator find nectar?

6. Where would a pollinator find pollen?

Pollination Syndrome B

Part 3: Think about possible pollinators.

Think about how an animal or insect pollinator might interact with your flower.

7. What characteristics might a *pollinator* have that would affect its ability to pollinate your flower?

Part 4: Use the "Flowers and Pollinators" resource.

8. Look at the tables in the resource. List your flower's characteristics below. On the right-hand side, list what kinds of pollinators might be attracted to the flower, based on the characteristics (there may be more than one possible pollinator).

Flower characteristic	Pollinator(s)
Shape/size	
Color	
Scent	
Food	
Day/night timing	

Mendel's Experiments

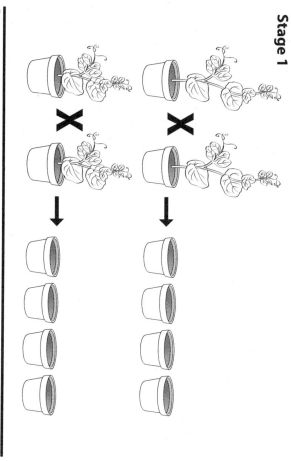
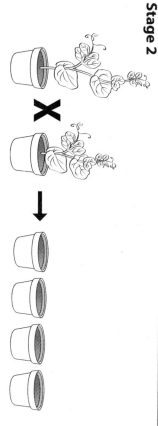
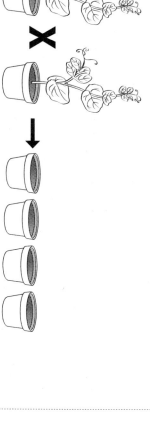

Mendel's Experiments

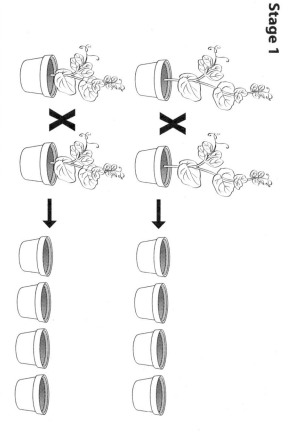
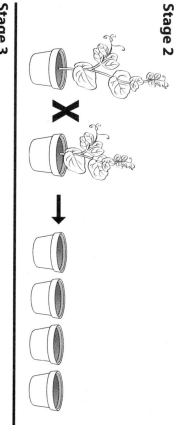
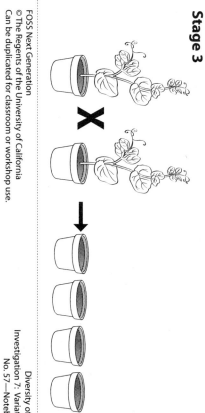

Pea Genotype to Phenotype

	Genotype
Stem-length alleles	ll
Flower-color alleles	F f

Pea plant genetic code

		From sperm ♂
Feature: Stem length **Traits** LL or Ll = tall stems ll = short stems		l
Feature: Flower color **Traits:** FF or Ff = purple flowers ff = white flowers		F

From egg ♀
l
f

Pea Genotype to Phenotype

	Genotype
Stem-length alleles	ll
Flower-color alleles	F f

Pea plant genetic code

		From sperm ♂
Feature: Stem length **Traits** LL or Ll = tall stems ll = short stems		l
Feature: Flower color **Traits:** FF or Ff = purple flowers ff = white flowers		F

From egg ♀
l
f

"Mendel and Punnett Squares" Question

Mendel's experiments showed that the feature of pea plant flower color was genetically determined. The dominant allele produced purple flowers (F). The recessive allele produced white flowers (f). Mendel crossed two pea plants, both with purple flowers. Some offspring had purple flowers and some had white flowers. Explain the result. Use a model to support your explanation.

Mendel's experiments showed that the feature of pea plant flower color was genetically determined. The dominant allele produced purple flowers (F). The recessive allele produced white flowers (f). Mendel crossed two pea plants, both with purple flowers. Some offspring had purple flowers and some had white flowers. Explain the result. Use a model to support your explanation.

"Mendel and Punnett Squares" Question

Mendel's experiments showed that the feature of pea plant flower color was genetically determined. The dominant allele produced purple flowers (F). The recessive allele produced white flowers (f). Mendel crossed two pea plants, both with purple flowers. Some offspring had purple flowers and some had white flowers. Explain the result. Use a model to support your explanation.

Mendel's experiments showed that the feature of pea plant flower color was genetically determined. The dominant allele produced purple flowers (F). The recessive allele produced white flowers (f). Mendel crossed two pea plants, both with purple flowers. Some offspring had purple flowers and some had white flowers. Explain the result. Use a model to support your explanation.

Hamster Parents

Pepper ♀

Female parent alleles		
	f	f

Sandy ♂

Male parent alleles		
	F	F

Hamster genetic code

Feature	Genotype	Trait
Fur color	F F or F f	Light fur
	f f	Dark fur

Hamster Parents

Pepper ♀

Female parent alleles		
	f	f

Sandy ♂

Male parent alleles		
	F	F

Hamster genetic code

Feature	Genotype	Trait
Fur color	F F or F f	Light fur
	f f	Dark fur

Insect Observations A

Part 1: General insect structure
Read introduction to "Insect Structure and Function."

1. Make three drawings of the hissing cockroach: one from the side, one from the top, and one from the front. Label the head, thorax, and abdomen in your drawings. Label the drawing as male or female.

2. Observe the hissing cockroach for several minutes. Describe what behaviors you observe.

3. Label the compound eyes on one of your drawings.

Part 2: Head (eyes, antennae, and mouthparts)
Read about insect heads.

4. What type of antenna do the cockroaches have? Circle below.

5. Hide a piece of food near the cockroach and observe its antennae. What do they do? What is the function of the antennae?

6. Use a toothpick to very carefully place a tiny bit of honey or syrup on one of the cockroach's antennae. What does the cockroach do? Why do you think this behavior is important?

7. What type of mouthparts does the cockroach have? Circle below.

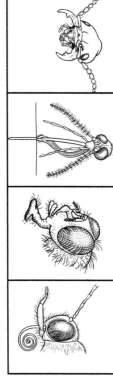

8. What kind of food do you think the cockroach eats? Put two or three different kinds of food in one of the dishes. Describe what you observe. Remove the food after your observations.

Insect Observations A

Part 1: General insect structure
Read introduction to "Insect Structure and Function."

1. Make three drawings of the hissing cockroach: one from the side, one from the top, and one from the front. Label the head, thorax, and abdomen in your drawings. Label the drawing as male or female.

2. Observe the hissing cockroach for several minutes. Describe what behaviors you observe.

3. Label the compound eyes on one of your drawings.

Part 2: Head (eyes, antennae, and mouthparts)
Read about insect heads.

4. What type of antenna do the cockroaches have? Circle below.

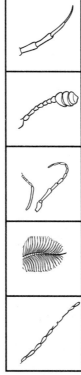

5. Hide a piece of food near the cockroach and observe its antennae. What do they do? What is the function of the antennae?

6. Use a toothpick to very carefully place a tiny bit of honey or syrup on one of the cockroach's antennae. What does the cockroach do? Why do you think this behavior is important?

7. What type of mouthparts does the cockroach have? Circle below.

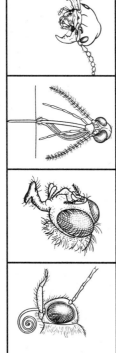

8. What kind of food do you think the cockroach eats? Put two or three different kinds of food in one of the dishes. Describe what you observe. Remove the food after your observations.

Insect Observations B

Part 3: Thorax (wings and legs)
Read about insect thoraxes.

9. Look for wings on the cockroach. Circle the type of wings it has.

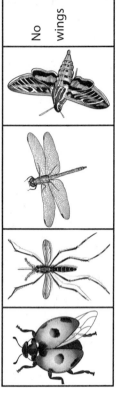

| | | | | No wings |

10. What does this tell you about the lifestyle of the cockroach?

11. Describe how the cockroach moves.

12. Circle the kind of legs the cockroach has.

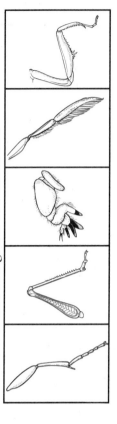

13. What part of the cockroach are the wings and legs attached to?

Part 4: Abdomen
Read Section 4 of "Insect Structure and Function."

14. What is contained in the abdomen?

15. What are the functions of those structures?

Part 5: Behavior

16. Note the fourth segment on the abdomen of the cockroach. Can you notice the spiracles? Why do cockroaches hiss?

17. What questions do you have about the Madagascar hissing cockroach? List at least two.

Structure/Behavior/Function Summary

Structure or behavior	Function(s)
Compound eye	
(head with antennae)	
(antenna)	
(leg)	
Hissing	
Pulling antenna through mouth	
Choosing dark damp places	

Structure/Behavior/Function Summary

Structure or behavior	Function(s)
Compound eye	
(head with antennae)	
(antenna)	
(leg)	
Hissing	
Pulling antenna through mouth	
Choosing dark damp places	

Comparing Systems

1. What is the transport system in each kind of organism?

    Vascular plants: _____

    Humans: _____

    Insects: _____

2. What is the function of each system?

    Vascular plants:

    Humans:

    Insects:

3. Compare the human and insect transport systems. How are they different?

4. Compare the organs of each system. How are they alike and how are they different? Make a table in your science notebook.

5. Compare the tissues of the human cardiovascular system and the insect circulatory system. How are they alike and how are they different? Make a table in your science notebook.

Bioblitz Summary and Reflections

How many different kinds of organisms did you predict and how many did your class actually find in the study site? Fill in the table.

Organism	Your prediction	Class count
Plants		
Fungi		
Lichens		
Animals collected		
Animal observations		

Answer these questions in your notebook.

1. Were the final numbers close to what you predicted? Explain.

2. Which organisms seem to be most abundant in the study site? Which organisms seem to be least abundant?

3. Do you think the class collection accurately represents the diversity of this site? Explain.

4. What was one thing you learned from this experience?

5. What questions remain?

Are Viruses Living Organisms?

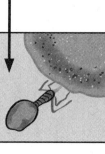

Bacteriophage T4 virus infecting an E. coli bacterium (bacterium is 200 nm long)

	Virus	Cell
Needs energy		✓
Needs water		✓
Grows		✓ (In multicellular organisms, cells increase in number, as well.)
Reproduces		Asexual or sexual reproduction; DNA is genetic material.
Needs suitable environment		✓
Responds to environment		✓
Exchanges gases		✓
Eliminates waste		✓
Structure		Cell structures and organelles.
Changes over time (evolves)		✓

In your notebook, write your conclusion. Are viruses living organisms? What is your evidence? If you cannot make a decision, what other information do you still need?

Tree of Life

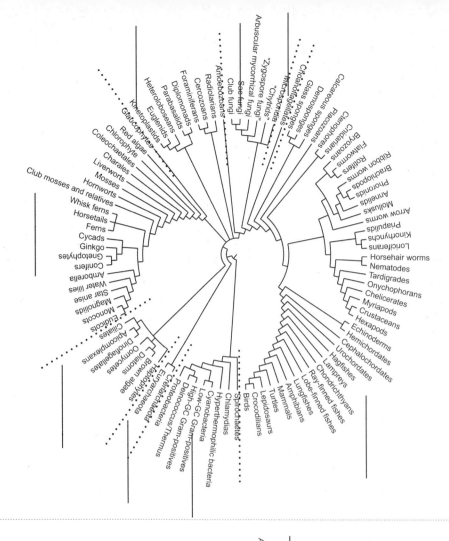

This version of the tree of life is based on data in *Life: The Science of Biology*, 9th ed., by D. Sadava, D. M. Hillis, H. C. Heller, and M. R. Berenbaum (Sunderland, MA: Sinauer Associates, 2011).

Tree of Life

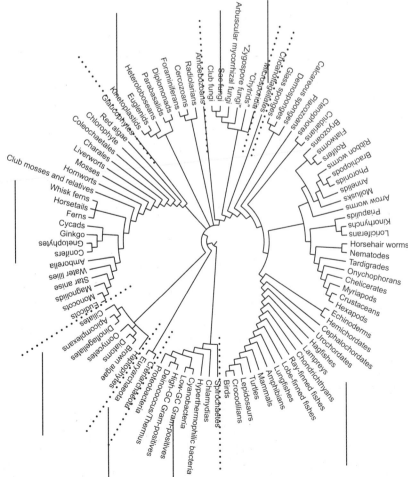

This version of the tree of life is based on data in *Life: The Science of Biology*, 9th ed., by D. Sadava, D. M. Hillis, H. C. Heller, and M. R. Berenbaum (Sunderland, MA: Sinauer Associates, 2011).

Teacher Masters

Teacher Master A

LIVING/NONLIVING CATEGORY TITLES

Living

Nonliving

Undecided

FOSS Next Generation
© The Regents of the University of California
Can be duplicated for classroom or workshop use.

Diversity of Life Course
Investigation 1: What Is Life?
Teacher Master A

MINIHABITAT SETUP

a. Place one spoon of soil in the bottom of a half-liter container.

b. Add dry leaves, dead grass, and twigs. The container should be no more than one-fourth full.

c. Add water. The container should be no more than half full. Leave some material sticking out of the water.

d. Put a lid on the container.

e. Label the container with your period, group number, and date.

WARNING — This set contains chemicals that may be harmful if misused. Read cautions on individual containers carefully. Not to be used by children except under adult supervision.

Teacher Master C

IS ANYTHING ALIVE IN HERE?

Materials

- 5 Vials with caps
- 1 Vial holder
- 6 Labels or 1 permanent marking pen
- 1 Cotton ball
- 5 Bags of unknown materials, labeled A–E

Part 1. Prepare the vials.

1. Use the labels or permanent marking pen to label the vials "A" through "E." Write your group number on each vial.
2. Put a label on the vial holder and write the date, your group number, the period, and the number of the liquid you have been assigned.
3. Pull the cotton ball apart and place the halves in vials A and D.

Part 2. Provide the liquid environment.

Add the liquid *assigned to your group* to the bottles as follows:

Vial A: 3 *full droppers* of liquid (*not* 3 drops)
Vial B: 30 mL of liquid
Vial C: 30 mL of liquid
Vial D: 3 *full droppers* of liquid (*not* 3 drops)
Vial E: 30 mL of liquid

Part 3. Add the unknown materials.

Safety note: Be careful not to mix the samples or touch them with your fingers. This could affect their survival if they are living organisms.

1. Measure 1 minispoon of material B into vial B. Measure 1 minispoon of material E into vial E. Put 8–10 grains of C into vial C. Cap and gently swirl the vials; do not shake them.
2. Sprinkle 1 minispoon of material A onto the damp cotton in vial A. Put 8–10 grains of material D on the damp cotton in vial D. Cap the vials.
3. Place the vials in the vial holder and return all other materials to the materials station.
4. After approximately 10 minutes, record any changes you observe on the *Five-Materials Observation* sheet. Also make drawings to show these changes.

WARNING — This set contains chemicals that may be harmful if misused. Read cautions on individual containers carefully. Not to be used by children except under adult supervision.

Teacher Master D

LIQUID-STATIONS INSTRUCTIONS

LIQUID 1

Vials A & D
- Put in half a cotton ball.
- Add 3 **droppers** of liquid 1 (not 3 drops).

Vials B, C, & E
- Add 30 mL of liquid 1.

LIQUID 2

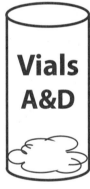

Vials A & D
- Put in half a cotton ball.
- Add 3 **droppers** of liquid 2 (not 3 drops).

Vials B, C, & E
- Add 30 mL of liquid 2.

LIQUID 3

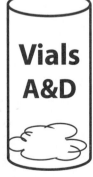

Vials A & D
- Put in half a cotton ball.
- Add 3 **droppers** of liquid 3 (not 3 drops).

Vials B, C, & E
- Add 30 mL of liquid 3.

FOSS Next Generation
© The Regents of the University of California
Can be duplicated for classroom or workshop use.

Diversity of Life Course
Investigation 1: What Is Life?
Teacher Master D

MICROSCOPE KIT *seed*

Teacher Master E

seed	seed	seed	seed	seed	seed	seed
seed	seed	seed	seed	seed	seed	seed
seed	seed	seed	seed	seed	seed	seed
seed	seed	seed	seed	seed	seed	seed
seed	seed	seed	seed	seed	seed	seed
seed	seed	seed	seed	seed	seed	seed
seed	seed	seed	seed	seed	seed	seed
seed	seed	seed	seed	seed	seed	seed
seed	seed	seed	seed	seed	seed	seed
seed	seed	seed	seed	seed	seed	seed
seed	seed	seed	seed	seed	seed	seed
seed	seed	seed	seed	seed	seed	seed
seed	seed	seed	seed	seed	seed	seed
seed	seed	seed	seed	seed	seed	seed

THE LETTER e

Teacher Master F

1.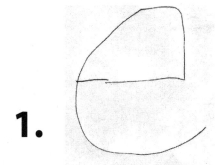

2.

3.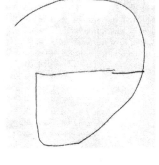

4.

5.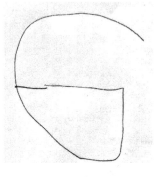

6.

FOSS Next Generation
© The Regents of the University of California
Can be duplicated for classroom or workshop use.

Diversity of Life Course
Investigation 2: The Microscope
Teacher Master F

HOW BIG IS IT? A

Teacher Master G

Courtesy of Alan Gould

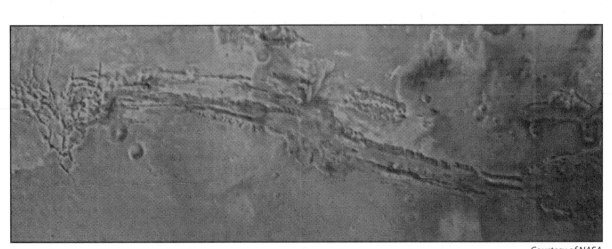

Courtesy of NASA

HOW BIG IS IT? B

Teacher Master H

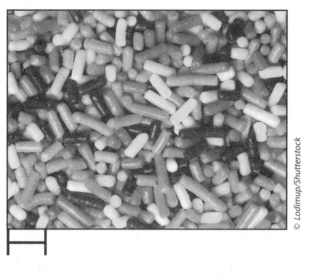

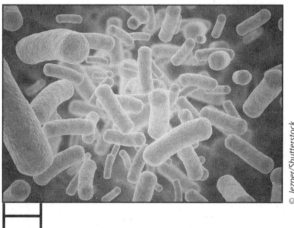

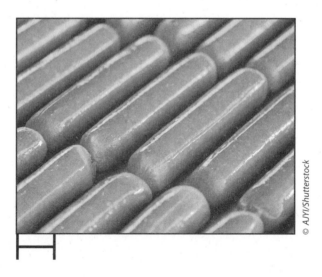

HOW BIG IS IT? C

Teacher Master I

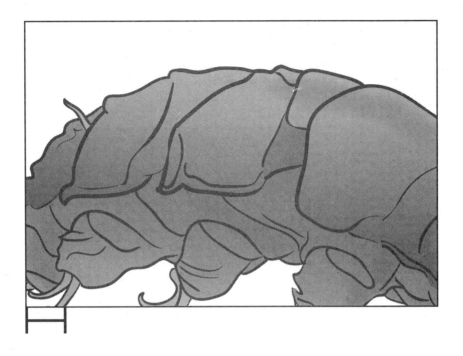

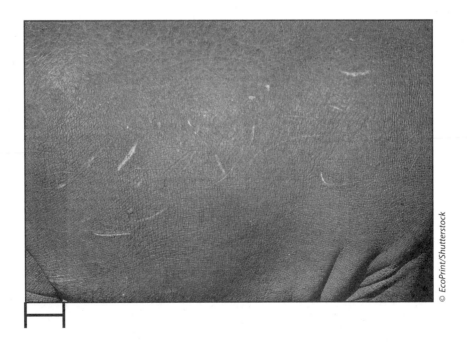

FOSS Next Generation
© The Regents of the University of California
Can be duplicated for classroom or workshop use.

Diversity of Life Course
Investigation 2: The Microscope
Teacher Master I

FIELD-OF-VIEW DIAMETER TABLES

Teacher Master J

Objective power	Magnification	Field of view
4X	40X	
10X	100X	
40X	400X	

Objective power	Magnification	Field of view
4X	40X	
10X	100X	
40X	400X	

Objective power	Magnification	Field of view
4X	40X	
10X	100X	
40X	400X	

Objective power	Magnification	Field of view
4X	40X	
10X	100X	
40X	400X	

Objective power	Magnification	Field of view
4X	40X	
10X	100X	
40X	400X	

Objective power	Magnification	Field of view
4X	40X	
10X	100X	
40X	400X	

Objective power	Magnification	Field of view
4X	40X	
10X	100X	
40X	400X	

Objective power	Magnification	Field of view
4X	40X	
10X	100X	
40X	400X	

Objective power	Magnification	Field of view
4X	40X	
10X	100X	
40X	400X	

Objective power	Magnification	Field of view
4X	40X	
10X	100X	
40X	400X	

Objective power	Magnification	Field of view
4x	40X	
10x	100X	
40x	400X	

Objective power	Magnification	Field of view
4x	40X	
10x	100X	
40x	400X	

EVIDENCE OF LIFE

Organism	Needs energy (food)	Needs water	Grows	Reproduces	Needs a suitable environment	Responds to environment	Exchanges gases	Eliminates waste

ELODEA LEAF LAYERS AND CELL SIZE

Teacher Master L

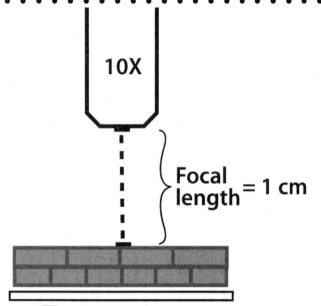

Focus is on top layer of leaf.

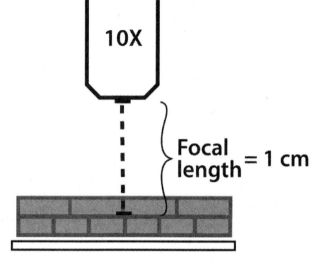

Focus is on bottom layer of leaf.

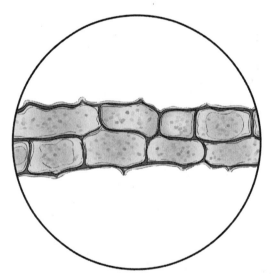

Elodea cells at 400X

CORK CELLS

Teacher Master M

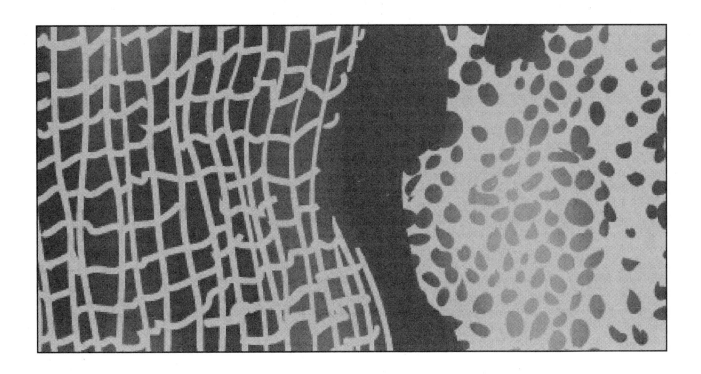

A view of cork cells based on Robert Hooke's observations in 1665

PLANT CELL STRUCTURES AND FUNCTIONS

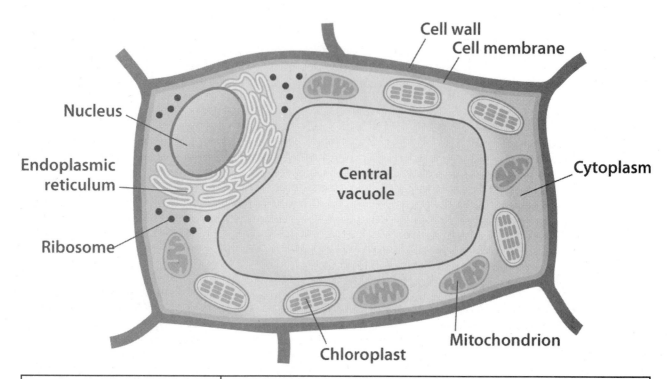

Cell structure	Function
Cell membrane	Boundary that controls what enters and leaves the cell
Cell wall	A rigid layer that supports the cell and provides shape
Central vacuole	A membrane that stores water and other substances, and provides structure and support for the cell.
Chloroplast	An organelle that converts the Sun's energy into food (sugars)
Cytoplasm	Internal fluid that contains the cell structures
Endoplasmic reticulum	A membranous structure that assembles proteins and parts of the cell membrane
Mitochondrion	An organelle that converts the energy in food into usable energy for the cell
Nucleus	Contains the cell's genetic material (DNA), which determines the nature of cell structures and substances
Ribosome	A structure that makes proteins, either free or bound to the endoplasmic reticulum

ELODEA AND PARAMECIUM

Teacher Master O

© Lawrence Hall of Science

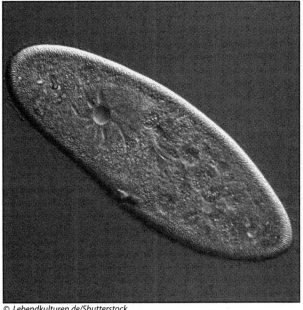
© Lebendkulturen.de/Shutterstock

1. Can a single living cell be a living organism?

2. Can a cell be living but not be an organism?

3. Is an elodea leaf cell an organism?

4. Is a paramecium cell an organism?

5. Is an elodea leaf an organism?

PROTIST CELL STRUCTURES AND FUNCTIONS

Teacher Master P

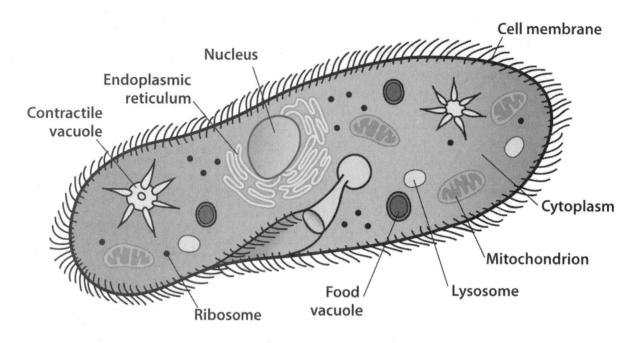

Cell structure	Function
Cell membrane	Boundary that controls what enters and leaves the cell
Contractile vacuole	A membrane that stores water and expels excess water
Cytoplasm	Internal fluid that contains the cell structures
Endoplasmic reticulum	A membranous structure that assembles proteins and parts of the cell membrane
Food vacuole	A membrane that stores food and merges with a lysosome to digest food
Lysosome	An organelle that digests cellular waste and merges with a food vacuole to digest food
Mitochondrion	An organelle that converts the energy in food into usable energy for the cell
Nucleus	An organelle that contains the cell's genetic material (DNA), which determines the nature of cell structures and substances
Ribosome	A structure that makes proteins, free or bound to the endoplasmic reticulum

FOSS Next Generation
© The Regents of the University of California
Can be duplicated for classroom or workshop use.

Diversity of Life Course
Investigation 3: The Cell
Teacher Master P

ANIMAL CELL STRUCTURES AND FUNCTIONS

Teacher Master Q

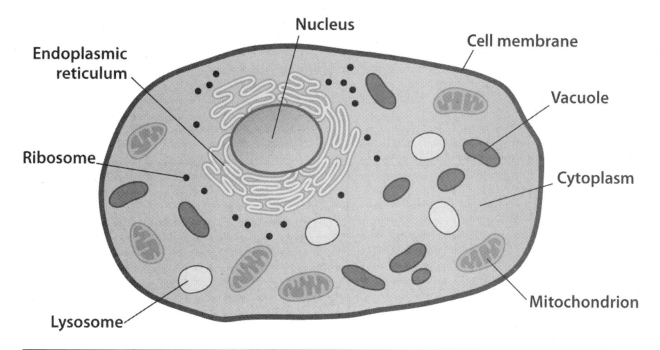

Cell structure	Function
Cell membrane	Boundary that controls what enters and leaves the cell
Cytoplasm	Internal fluid that contains the cell structures
Endoplasmic reticulum	A membranous structure that assembles proteins and parts of the cell membrane
Lysosome	An organelle that digests cellular waste
Mitochondrion	An organelle that converts the energy in food into usable energy for the cell
Nucleus	An organelle that contains the cell's genetic material (DNA), which determines the nature of cell structures and substances
Ribosome	A structure that makes proteins, either free or bound to the surface of the endoplasmic reticulum
Vacuole	A membrane stores water and other materials

CELL TRACKING

Teacher Master R

Structure	Plant cell	Protist cell (paramecia)	Animal cell	Bacterial cell	Fungal cell	Archaean cell
Cell membrane						
Cell wall						
Chloroplasts						
Cytoplasm						
Endoplasmic reticulum						
Lysosomes						
Mitochondria						
Nucleus						
Plasmids						
Ribosomes						
Vacuoles						

SIMILARITY

Teacher Master S

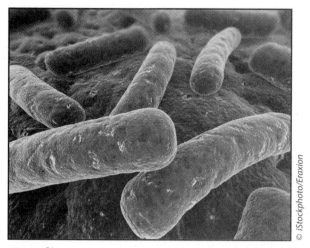
E. coli

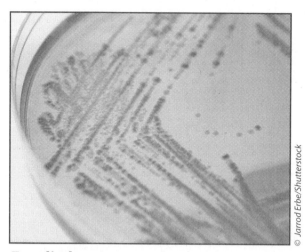
E. coli plate

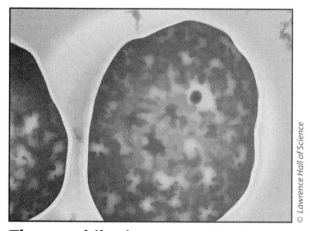
Thermophiles in Yellowstone

Yellowstone hot springs

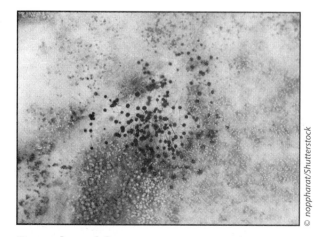
Bread mold

Bread mold

INOCULATING AN AGAR PLATE

Teacher Master T

1. Work with your group. Select an object, material, or location to test for bacteria.

2. Collect a sample. For surfaces or liquids, use a cotton swab.

   - For surfaces, rub and roll the cotton end of the swab on the surface.

   - For liquids, touch the swab in the liquid to get liquid on the swab.

   NOTE: Don't let the swab touch anything except what you are testing.

3. Inoculate the sterile agar. Lift one edge of the petri-dish lid just high enough to insert the swab. Make an S streak across the surface of the agar in your quarter of the petri dish. Each member marks a different quarter of the dish.

4. Tape the lid onto the bottom of the dish, using two small pieces of tape. Label the petri dish with your group number and class period. Record on your notebook sheet what you used to inoculate each section of the dish. **You will not open the dish again!**

5. Place the petri dish in a warm, dark place. Store it upside down so that any moisture inside the dish will not drip onto the agar.

UNITS

Teacher Master U

Length	Unit	Symbol
1 m	meter	m
1/100 of a m	centimeter	cm
1/1,000 of a m	millimeter	mm
1/1,000 of a mm	micrometer	μm
1/1,000 of a μm	nanometer	nm
1/10 of a nm (1/10 billionth of a m)	angstrom	Å

Length	Unit	Symbol
1 m	meter	m
1/100 of a m	centimeter	cm
1/1,000 of a m	millimeter	mm
1/1,000 of a mm	micrometer	μm
1/1,000 of a μm	nanometer	nm
1/10 of a nm (1/10 billionth of a m)	angstrom	Å

Length	Unit	Symbol
1 m	meter	m
1/100 of a m	centimeter	cm
1/1,000 of a m	millimeter	mm
1/1,000 of a mm	micrometer	μm
1/1,000 of a μm	nanometer	nm
1/10 of a nm (1/10 billionth of a m)	angstrom	Å

Length	Unit	Symbol
1 m	meter	m
1/100 of a m	centimeter	cm
1/1,000 of a m	millimeter	mm
1/1,000 of a mm	micrometer	μm
1/1,000 of a μm	nanometer	nm
1/10 of a nm (1/10 billionth of a m)	angstrom	Å

Length	Unit	Symbol
1 m	meter	m
1/100 of a m	centimeter	cm
1/1,000 of a m	millimeter	mm
1/1,000 of a mm	micrometer	μm
1/1,000 of a μm	nanometer	nm
1/10 of a nm (1/10 billionth of a m)	angstrom	Å

Length	Unit	Symbol
1 m	meter	m
1/100 of a m	centimeter	cm
1/1,000 of a m	millimeter	mm
1/1,000 of a mm	micrometer	μm
1/1,000 of a μm	nanometer	nm
1/10 of a nm (1/10 billionth of a m)	angstrom	Å

Length	Unit	Symbol
1 m	meter	m
1/100 of a m	centimeter	cm
1/1,000 of a m	millimeter	mm
1/1,000 of a mm	micrometer	μm
1/1,000 of a μm	nanometer	nm
1/10 of a nm (1/10 billionth of a m)	angstrom	Å

Length	Unit	Symbol
1 m	meter	m
1/100 of a m	centimeter	cm
1/1,000 of a m	millimeter	mm
1/1,000 of a mm	micrometer	μm
1/1,000 of a μm	nanometer	nm
1/10 of a nm (1/10 billionth of a m)	angstrom	Å

Length	Unit	Symbol
1 m	meter	m
1/100 of a m	centimeter	cm
1/1,000 of a m	millimeter	mm
1/1,000 of a mm	micrometer	μm
1/1,000 of a μm	nanometer	nm
1/10 of a nm (1/10 billionth of a m)	angstrom	Å

Length	Unit	Symbol
1 m	meter	m
1/100 of a m	centimeter	cm
1/1,000 of a m	millimeter	mm
1/1,000 of a mm	micrometer	μm
1/1,000 of a μm	nanometer	nm
1/10 of a nm (1/10 billionth of a m)	angstrom	Å

Length	Unit	Symbol
1 m	meter	m
1/100 of a m	centimeter	cm
1/1,000 of a m	millimeter	mm
1/1,000 of a mm	micrometer	μm
1/1,000 of a μm	nanometer	nm
1/10 of a nm (1/10 billionth of a m)	angstrom	Å

Length	Unit	Symbol
1 m	meter	m
1/100 of a m	centimeter	cm
1/1,000 of a m	millimeter	mm
1/1,000 of a mm	micrometer	μm
1/1,000 of a μm	nanometer	nm
1/10 of a nm (1/10 billionth of a m)	angstrom	Å

FOSS Next Generation
© The Regents of the University of California
Can be duplicated for classroom or workshop use.

Diversity of Life Course
Investigation 4: Domains
Teacher Master U

LEVELS OF COMPLEXITY

Teacher Master V

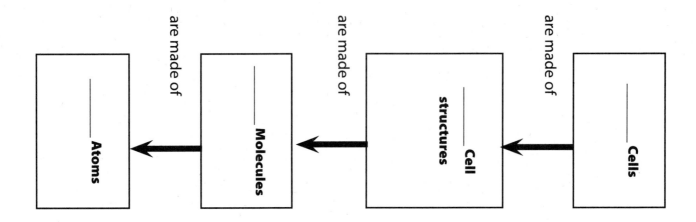

OR

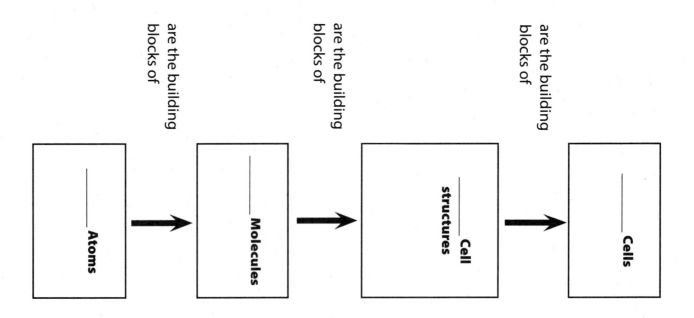

BACTERIAL CELL STRUCTURES AND FUNCTIONS

Teacher Master W

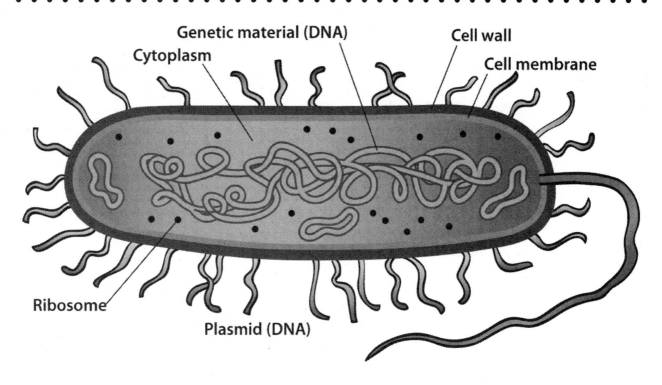

Cell structure	Function
Cell membrane	Boundary that controls what enters and leaves the cell
Cell wall	A rigid layer that supports the cell and provides shape
Cytoplasm	Internal fluid that contains the cell structures
Genetic material (DNA)	A molecule that determines the nature of cell structures and substances
Plasmid	Small piece of genetic material that is independent of other DNA in the cell and that can be passed to other bacteria
Ribosome	A structure that makes proteins, in the cytoplasm

DOTS

Teacher Master X

ONE MILLION DOLLARS

Day 1: $0.01
Day 2: $0.02
Day 3: $0.04
Day 4: $0.08
Day 5: $0.16
Day 6: $0.32
Day 7: $0.64
Day 8: $1.28
Day 9: $2.56
Day 10: $5.12
Day 11: $10.24
Day 12: $20.48
Day 13: $40.96
Day 14: $81.92
Day 15: $163.84
Day 16: $327.68
Day 17: $655.36
Day 18: $1,310.72
Day 19: $2,621.44
Day 20: $5,242.88
Day 21: $10,485.76
Day 22: $20,971.52
Day 23: $41,943.04
Day 24: $83,886.08
Day 25: $167,772.16
Day 26: $335,544.32
Day 27: $671,088.64
Day 28: $1,342,177.28
Day 29: $2,684,354.56
Day 30: $5,368,709.12

EXPONENTIAL GROWTH

Teacher Master Z

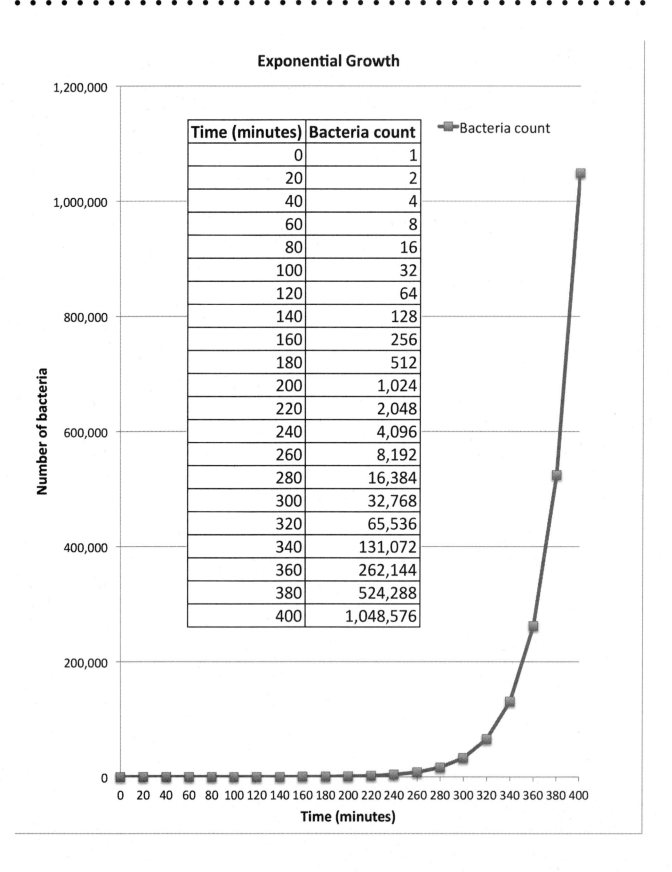

Exponential Growth

Time (minutes)	Bacteria count
0	1
20	2
40	4
60	8
80	16
100	32
120	64
140	128
160	256
180	512
200	1,024
220	2,048
240	4,096
260	8,192
280	16,384
300	32,768
320	65,536
340	131,072
360	262,144
380	524,288
400	1,048,576

FUNGAL CELL STRUCTURES AND FUNCTIONS

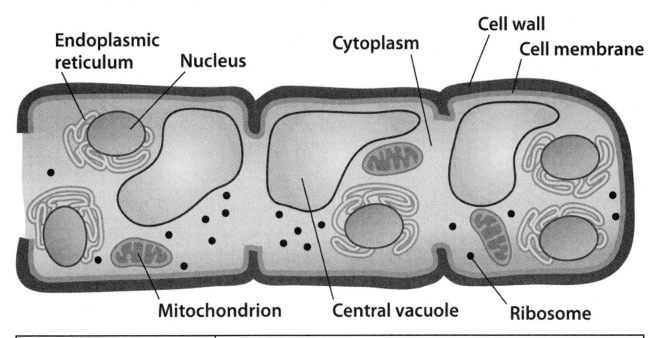

Cell structure	Function
Cell membrane	Boundary that controls what enters and leaves the cell
Cell wall	A rigid layer that supports the cell and provides shape
Central vacuole	A membrane that stores water and other substances, and provides structure and support for the cell
Cytoplasm	Internal fluid that contains the cell structures
Endoplasmic reticulum	A membranous structure that assembles proteins and parts of the cell membrane
Mitochondrion	An organelle that converts the energy in food into usable energy for the cell
Nucleus	An organelle that contains the cell's genetic material (DNA), which determines the nature of cell structures and substances
Ribosome	A structure that makes proteins, either free or bound to the endoplasmic reticulum

Teacher Master BB

CELERY-INVESTIGATION CONSIDERATIONS

Materials

1	Stalk of celery
1	Vial
1	Vial holder
1	Syringe
1–2	Accurate balances (for class)
•	Water

Things to Consider

- How will you determine if water is lost to evaporation?

- How will you determine if water is absorbed into the celery?

- What tools will you use to make your measurements?

CELERY-INVESTIGATION CLASS RESULTS

Group	Water gone from celery vial (mL or g)	Change in mass of the celery (g)	Water unaccounted for (mL or g)

Teacher Master DD

SEED HUNT

Collect one or two of as many kinds of seeds as you can find. Look for seeds from grasses, trees, bushes, fruits, vegetables. Use transparent tape to tape the seeds on the sheet. Label each seed with where it came from and how you think it moves away from the parent plant (how the seed disperses).

1	Taken from: Dispersal method:	2	Taken from: Dispersal method:	3	Taken from: Dispersal method:	4	Taken from: Dispersal method:
5	Taken from: Dispersal method:	6	Taken from: Dispersal method:	7	Taken from: Dispersal method:	8	Taken from: Dispersal method:

FOSS Next Generation
© The Regents of the University of California
Can be duplicated for classroom or workshop use.

Diversity of Life Course
Investigation 6: Plant Reproduction and Growth
Teacher Master DD

GRAINS
Wheat

Barley

Oats

Corn

WARNING — This set contains chemicals that may be harmful if misused. Read cautions on individual containers carefully. Not to be used by children except under adult supervision.

Teacher Master FF

SALTWATER GERMINATION SETUP

Each group gets four petri dishes and two paper towels.

a. Fold each paper towel into quarters and use the smaller half of a petri dish to trace a circle on the top quarter of one of the folded towels. Cut all four layers of the paper towel at once by cutting on the circle line. Repeat so that you have eight circles.

b. Place two paper towel circles in the larger half of the petri dish. The smaller half of the dish will go on top as a cover.

c. Label both halves of the petri dish with self-stick notes: group name, period, number of spoons of salt, name of seeds.

d. Put 5 mL of the correct salt solution into the half with the paper towels. Spread it around so the entire paper-towel circle is wet.

e. Count 40 of your assigned seeds into a plastic cup. Put 10 seeds in each dish. Scatter them across the paper towel. Put the smaller half of the petri dish on top as a cover and make sure the labels are secure.

f. Put the dishes in the designated tub. The tub will be placed in a room-temperature dark location or covered with newspaper. This is day 0. On day 2, you will make your first observations and place the dishes in the light.

ROOTS AND SHOOTS

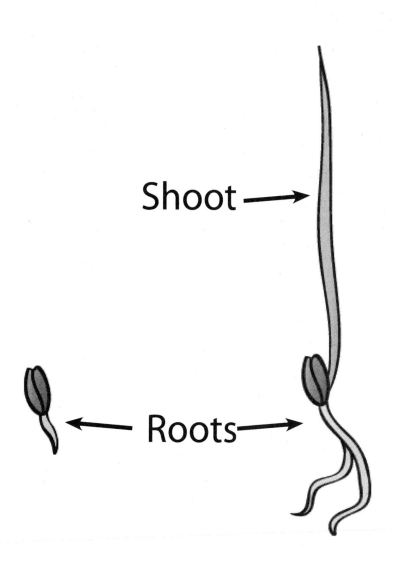

Teacher Master HH

WARNING — This set contains chemicals that may be harmful if misused. Read cautions on individual containers carefully. Not to be used by children except under adult supervision.

PLANT PROFILE

Follow the model below to set up four plant profiles, one for each petri dish. Place one strip of clear packing tape over the seeds. Place other strips over the shoots and roots if necessary.

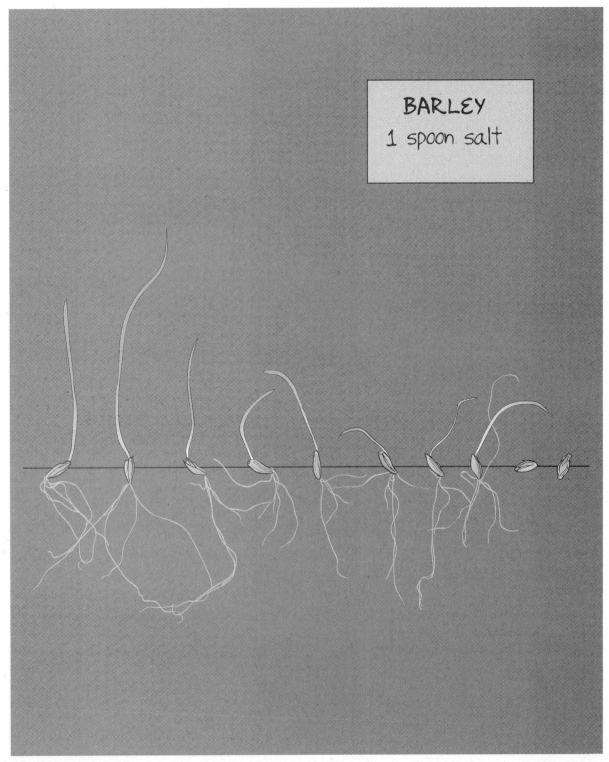

BARLEY
1 spoon salt

PLANT-REPRODUCTION SEQUENCE

J	Ovules, which contain the female egg cells, form in the ovary.
E	Pollen grains, which contain the male sperm cells, form on the anthers.
A	A pollen grain, usually carried by animal or air, lands on the stigma of another flower.
G	The pollen grain forms a long tube down the length of the pistil into the ovule.
C	A sperm cell travels down the pollen tube.
F	The sperm cell fertilizes an egg. The egg and sperm fuse to form a single cell with information from the male and female.
I	The single cell divides, and each of those cells divides, and so on until the many cells develop into an embryo.
B	The parent plant forms a food source for the developing embryo.
H	The seed-containing ovary develops into a fruit.
D	Fruit is dropped or consumed by an animal, and the seed is released.

BEE AND FLOWER

Teacher Master JJ

© Kovalchuk Oleksandr/Shutterstock

HUMAN FEATURES AND TRAITS

Teacher Master KK

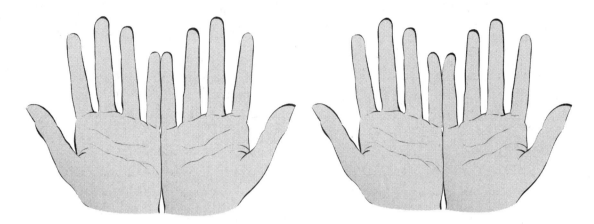

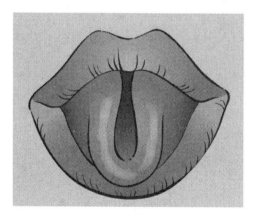

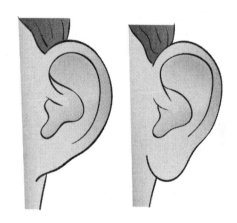

FOSS Next Generation
© The Regents of the University of California
Can be duplicated for classroom or workshop use.

Diversity of Life Course
Investigation 7: Variation of Traits
Teacher Master KK

FEATURES AND TRAITS

Teacher Master LL

Human features											
Little fingers			Ears			Tongue			Hairline		
Traits			Traits			Traits			Traits		
Straight	Bent	Can't tell	Free	Attached	Can't tell	Flat	Curled	Can't tell	Straight	Peak	Can't tell

PEA PLANTS

Teacher Master MM

MENDEL'S EXPERIMENT—GETTING READY

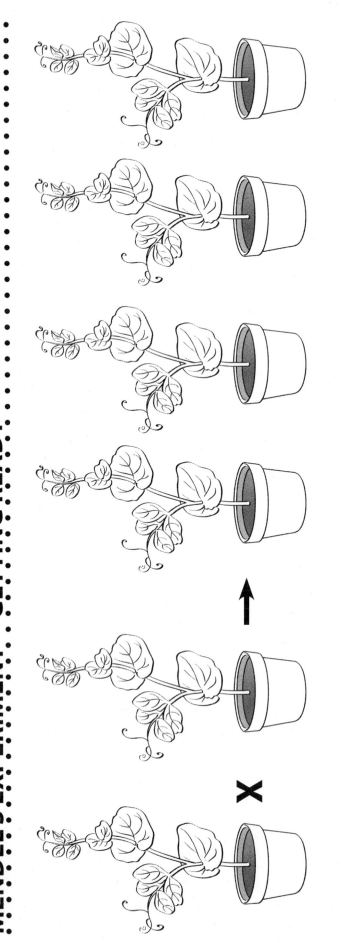

MENDEL'S EXPERIMENT—SHORT AND TALL PARENTS

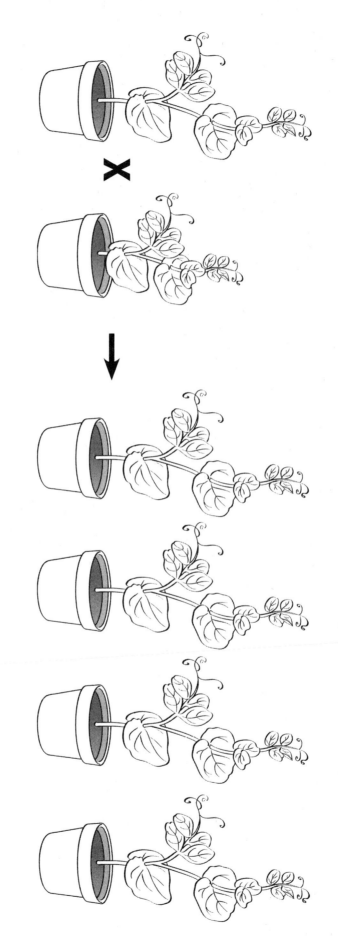

PARENTS

F₁ GENERATION

MENDEL'S EXPERIMENT—F₁ PARENTS

F₁ GENERATION

F₂ GENERATION

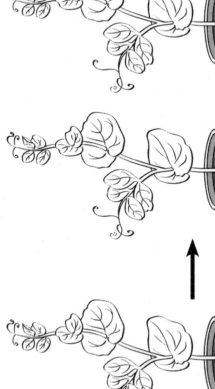

PEA PLANT ALLELES

Teacher Master QQ

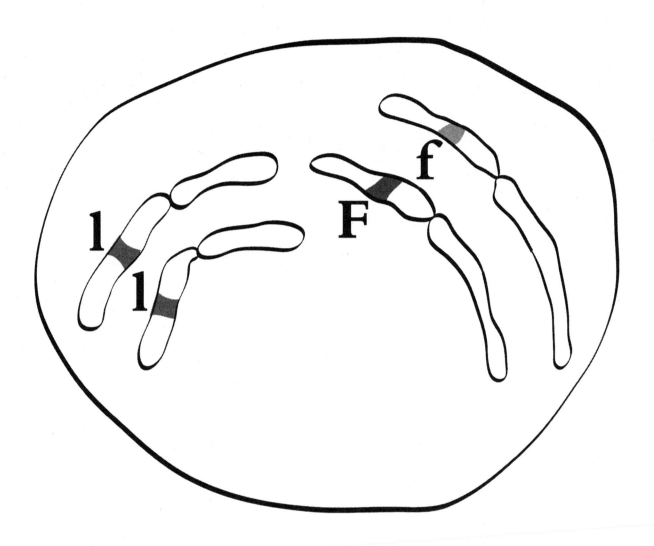

Pea plant alleles	
	Genotype
Stem-length alleles	l l
Flower-color alleles	F f

PEA GENOTYPE TO PHENOTYPE

Teacher Master RR

	Genotype
Stem-length alleles	l l
Flower-color alleles	F f

From egg ♀	Pea plant genetic code	From sperm ♂
l	**Feature: Stem length**	l
	Traits LL or Ll = tall stems; ll = short stems	
f	**Feature: Flower color**	F
	Traits FF or Ff = purple flowers; ff = white flowers	

HAMSTER PARENTS

Teacher Master SS

Pepper ♀

Female parent alleles	
f	f

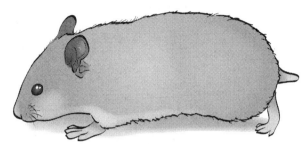

Sandy ♂

Male parent alleles	
F	F

Hamster genetic code		
Feature	Genotype	Trait
Fur color	F F or F f	Light fur
	f f	Dark fur

THE THREE DOMAINS

Teacher Master TT

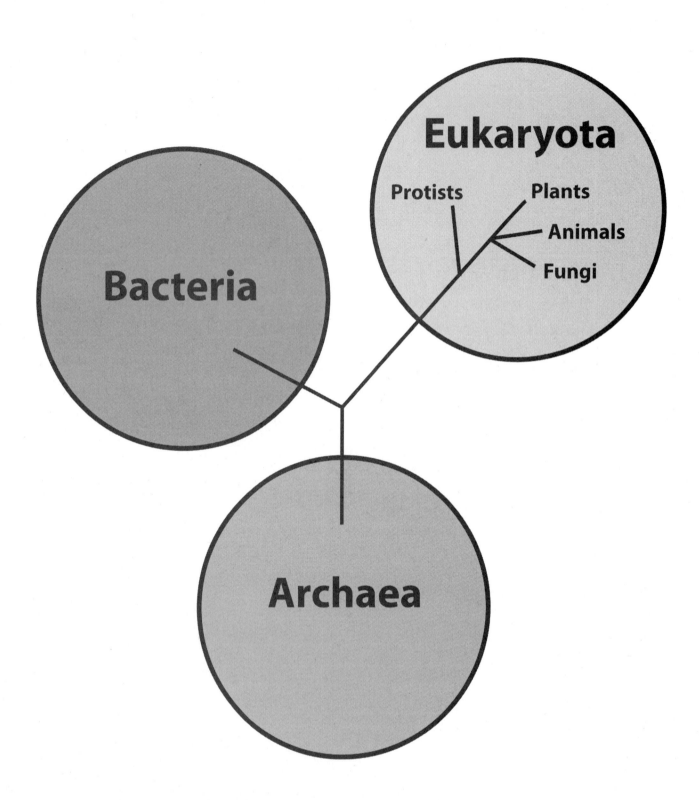

THE MADAGASCAR HISSING COCKROACH

Male

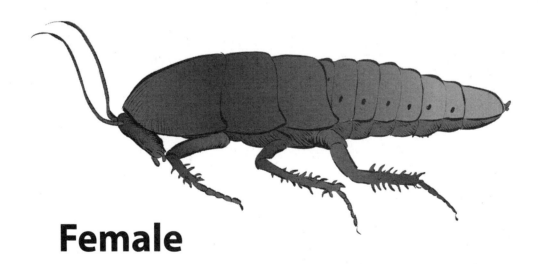

Female

SYSTEMS SUMMARY

Answer these questions in your group. Turn in one sheet with your group's responses.

1. How are multicellular organisms the same as and different than single-celled organisms? You can use a box and T-chart to organize your response.

2. How do the cells in a multicellular organism contribute to the survival of the organism? Give at least one example to support your response.

3. Are organs made up of just one kind of tissue? Give at least one example to support your response.

4. Why do you think it is important for multicellular organisms to have tissues, organs, and organ systems as opposed to just having different kinds of cells?

PLANT-DIVERSITY ACTION CARD

Your team will collect one leaf sample from as many different plants as you can find in your study site.

Things to Consider

- Do not collect flowers, seeds, and cones. They are good identifying structures for different kinds of plants, but we are only collecting leaf samples.

- Moss is a type of plant. If you find any, bring back a small sample.

- Lichen is a combination of fungi and algae that looks like a fuzzy cluster in trees or a spreading mat on rocks. If you find any, take only a small sample to identify it during our sort.

- Do **not** touch or collect any mushrooms or other fungi you find. Bring back a description detailed enough to distinguish it from observations other groups may make.

ANIMAL-DIVERSITY ACTION CARD

Teacher Master XX

Role	Description
Recorder	• Supervises the transfer of new animals into vials • Eliminates any repetition in collection • Adds new samples to data sheets; works with collector to come up with an effective name for animal • Makes sure collected organisms are out of the Sun
Collector	• Uses all techniques to collect as many different types of organisms as possible in the short time given • Coordinates with other collectors to minimize overlap • Brings samples to Recorder; transfers them into vials for identification

Collection Techniques

Subterranean. Lift rocks and logs away from you to collect animals underneath. Replace rocks and logs so that the area and animal homes are minimally disturbed.

Litter layer. Sift through soil and the layer of leaves and twigs to collect animals.

Grass and shrub layer. Open the zip bag and fold back the edges to use it like a sweep net. Brush it through shrubs or tall grass to collect animals. As soon as you stop moving the bag, close the opening to avoid losing any captured organisms.

Observations. Add observations of larger animals you see (or animal signs) to the observation list. Give each sign or animal a description to help distinguish it from other groups' observations (examples: small yellow bird, small brown squirrel, 2 cm wide tunnel opening).

BIOBLITZ DATA SHEET

Teacher Master YY

Organism	Class count
Plants	
Fungi	
Lichens	
Animals collected	
Animal observations	

Look through your collection. Create a descriptive name for each organism (examples: green hopper, red ant, long-legged spider.) Add other observations below.

	Descriptive name of animal
1	
2	
3	
4	
5	
6	
7	
8	
9	
10	

VIRUS: CLASS QUOTATIONS A

There is some debate about whether viruses themselves should be considered living . . . . The debate, in my opinion, is a . . . largely unimportant one. Viruses are completely dependent on other organisms for elements of their life cycle, but that is no different than the rest of known life forms, none of which, to my knowledge, could live in a world devoid of other life. Either way, it is clear that viruses are part of the living systems of our planet . . . .

Nathan Wolfe, *The Viral Storm* (New York: Henry Holt and Co., 2011)

• •

Viruses exist in two distinct states. When not in contact with a host cell, the virus remains entirely dormant. During this time there are no internal biological activities occurring within the virus . . . . In this simple, clearly non-living state viruses are referred to as "virions." Virions can remain in this dormant state for extended periods of time, waiting patiently to come into contact with the appropriate host. When the virion comes into contact with the appropriate host, it becomes active and is then referred to as a virus. It now displays properties [seen in] living organisms, such as reacting to its environment and directing its efforts toward [reproduction].

"An Introduction to the Bacteriophage T4 Virus," http://www.dform.com/projects/t4/virus.html

• •

A virus (from the Latin *virus* meaning toxin or poison) is a microscopic organism consisting of genetic material (RNA or DNA) surrounded by a protein, lipid (fat), or glycoprotein coat . . . . Viruses are unique organisms because they cannot reproduce without a host cell. After contacting a host cell, a virus will insert genetic material into the host and take over that host's functions. The cell, now infected, continues to reproduce, but it reproduces more viral protein and genetic material instead of its usual products.

Peter M. Crosta, "What Is a Virus? What Is a Viral Infection?" (2015)
http://www.medicalnewstoday.com/articles/158179.php

• •

When researchers first discovered agents that behaved like bacteria but were much smaller and caused diseases such as rabies and foot-and-mouth disease, it became the general view that viruses were biologically "alive." However, this perception changed in 1935 when the tobacco mosaic virus was crystallized and it was shown that the particles lacked the mechanisms necessary [to survive]. Once it was established that viruses consist merely of DNA or RNA surrounded by a protein shell, it became the scientific view that they are more [like] complex biochemical mechanisms than living organisms.

George Rice, "Are Viruses Alive?" http://serc.carleton.edu/microbelife/yellowstone/viruslive.html

VIRUS: CLASS QUOTATIONS B

Teacher Master AAA

Scientists have always recognized the importance of viruses, but recently it has become clearer that viruses are an integral [important] part of every ecosystem and can't be ignored when we try to understand how life on Earth works. We usually only hear about viruses in the context of human disease. But most viruses are actually not harmful, and in fact have played an important part in evolution and in maintaining healthy ecosystems.

Forest Rohwer, quoted in Hamish Clarke, "A Few Good Viruses,"
http://scidstuff.wordpress.com/2007/02/07/a-few-good-viruses/

Are viruses alive? Until very recently, answering this question was often negative and viruses were not considered in discussions on the origin and definition of life. This situation is rapidly changing, following several discoveries that have modified our vision of viruses. It has been recognized that viruses have played (and still play) a major innovative role in the evolution of cellular organisms. New definitions of viruses have been proposed and their position in the universal tree of life is actively discussed. Viruses . . . can be viewed as complex living entities [organisms] that transform the infected cell into a novel organism—the virus—producing [new viral particles]. . . . I propose to define an organism as an ensemble of integrated organs (molecular or cellular) producing [new] individuals.

Patrick Forterre, "Defining Life: The Virus Viewpoint,"
http://www.ncbi.nlm.nih.gov/pmc/articles/PMC2837877/

Cells of free-living organisms, including bacteria, contain a variety of organelles essential for life such as ribosomes that manufacture proteins, mitochondria or other structures that generate energy, and complex membranes for transporting molecules within the cell, and also across the cell wall. Viruses, not being cells, have none of these and are therefore inert [inactive] until they infect a living cell. Virus particles resemble seeds which can only spring into life when they find the right soil. But unlike seeds, viruses do not carry the genes to code for all the proteins they require to 'germinate' and complete their life cycle. So they hijack a cell's organelles and use what they need, often killing the cell in the process.

Dorothy H. Crawford, *Viruses: A Very Short Introduction* (Oxford: Oxford University Press, 2011)

For about 100 years, the scientific community has repeatedly changed its collective mind over what viruses are. First seen as poisons, then as life-forms, then biological chemicals, viruses today are thought of as being in a gray area between living and nonliving: they cannot replicate [reproduce] on their own but can do so in truly living cells and can also affect the behavior of their hosts profoundly. The categorization of viruses as nonliving during much of the modern era of biological science has had an unintended consequence: it has led most researchers to ignore viruses in the study of evolution. Finally, however, scientists are beginning to appreciate viruses as fundamental players in the history of life.

Luis P. Villarreal, "Are Viruses Alive?"
http://www.scientificamerican.com/article.cfm?id=are-viruses-alive-2004

VIRUS: FINAL QUOTATION

Teacher Master BBB

"Maybe the most important question is not whether viruses are alive, but rather how much of a role they play in the movement and molding of life as we recognize it today."

Adapted from George Rice, "Are Viruses Alive?," http://serc.carleton.edu/microbelife/yellowstone/viruslive.html

"Maybe the pertinent question is not whether viruses are alive, but rather to what extent do they play a role in the movement and molding of life as we perceive it today?"

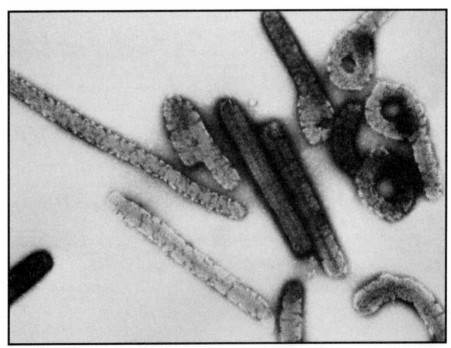

© CDC/Dr. Erskine Palmer, Russell Regnery, PhD

DIVERSITY OF LIFE KEY POINTS A

Teacher Master CCC

- Any free-living thing is an organism.
- ★ All organisms exhibit common characteristics and have certain requirements: they grow, need energy (food) and water, exchange gases, respond to the environment, reproduce, eliminate waste, and need a suitable environment in which to live.
- Something can be dead only if it was once living.
- Some organisms can become dormant to survive in an unsuitable environment.
- A compound optical microscope is composed of a two-lens system (eyepiece and objective lens), a stage on which to mount the material being observed, a light source (lamp or reflected), and a focusing system.
- A microscope's optical power is the product of the magnification of the eyepiece and the objective lens.
- The field of view is the diameter of the circle of light seen through the microscope. As the power increases, the field of view decreases.
- A microscope may reverse and invert images.
- ★ The cell is the basic unit of life. All living things are made up of one or more cells.
- ★ Every cell has structures that enable it to carry out life's functions.
- Both single-celled and multicellular organisms exhibit all the characteristics of life.
- ★ Asexual reproduction is a method of reproduction that results in offspring with identical genetic information.
- ★ Cells are made of cell structures, which are made of molecules, which are made of atoms.
- Bacteria, fungi, and archaea demonstrate all the characteristics of life.
- ★ Life is classified into three domains (Archaea, Bacteria, Eukaryota), depending upon cellular and molecular characteristics.
- ★ Transpiration is the process by which water is carried through vascular plants from the roots to stomata, ensuring that all the cells have access to water.
- The vascular system of plants consists of xylem and phloem.
- ★ Plants use photosynthesis and aerobic cellular respiration to make usable energy from the Sun's energy.
- ★ Cells are the building blocks of tissues, which are the building blocks of organs, which are the building blocks of organ systems, which are the building blocks of multicellular organisms.

★ These key points are big ideas introduced in the **Diversity of Life Course**.

DIVERSITY OF LIFE KEY POINTS B

- ★ Environmental and genetic factors affect the germination and growth of plants.
- ★ Flowering plants reproduce sexually, producing seeds, which contain dormant new plants.
- ★ Flowering plants have characteristics that attract pollinators to ensure successful pollination and reproduction.
- Pollinators are attracted to flowers that meet their needs.
- ★ Genetic information (DNA) is passed from parents to offspring in both sexual and asexual reproduction.
- Genes define an organism's genotype.
- Genes code for proteins, which determine an organism's phenotype (traits).
- In sexually reproducing organisms, each parent contributes half the offspring's alleles so that the offspring have genotypes that are not identical to either parent. In asexual reproducing organisms (protists and bacteria), the "mother" organism divides reproducing two daughter offspring with identical genes as the "mother" (parent).
- A Punnett square is a mathematical model used to predict the ratio of genotypes and phenotypes in future generations of sexually reproducing organisms.
- ★ The structures and behaviors of an organism have functions that enhance the organism's chances to survive and reproduce in its habitat.
- Insects have open circulatory systems that transport substances to and away from their cells.
- ★ Biodiversity is the variety of life that exists in a particular habitat or ecosystem.
- Measuring biodiversity includes measuring both the variety of organisms and the number of organisms in a habitat or ecosystem.
- Scientific debate regarding whether viruses are living organisms is ongoing.
- All life on Earth is related.

- ★ These key points are big ideas introduced in the **Diversity of Life Course**.

Assessment Masters

Embedded Assessment Notes

Diversity of Life

Investigation ___, Part ___ Date _____

Concept:

Tally: Got it _____ | Doesn't get it _____

Misconceptions/incomplete ideas:

Reflections/next steps:

Investigation ___, Part ___ Date _____

Concept:

Tally: Got it _____ | Doesn't get it _____

Misconceptions/incomplete ideas:

Reflections/next steps:

Investigation ___, Part ___ Date _____

Concept:

Tally: Got it _____ | Doesn't get it _____

Misconceptions/incomplete ideas:

Reflections/next steps:

Performance Assessment Checklist by Group

Diversity of Life

Investigation 2, Part 1

Group	Science and Engineering Practices		Crosscutting Concepts
	Planning and carrying out investigations	Using mathematics and computational thinking	Scale, proportion, and quantity

NOTE: See the Assessment chapter for a discussion about how to use this checklist.

Performance Assessment Checklist by Student

Diversity of Life

Investigation 2, Part 1

Student	Science and Engineering Practices		Crosscutting Concepts
	Planning and carrying out investigations	Using mathematics and computational thinking	Scale, proportion, and quantity

NOTE: See the Assessment chapter for a discussion about how to use this checklist.

Performance Assessment Checklist by Group

Diversity of Life

Investigation 2, Part 3

Group	Science and Engineering Practices		DCI	Crosscutting Concepts	
	Planning and carrying out investigations	Using mathematics and computational thinking	LS1.A Structure and function	Scale, proportion, and quantity	Structure and function

FOSS Next Generation
© The Regents of the University of California
Can be duplicated for classroom or workshop use.

NOTE: See the Assessment chapter for a discussion about how to use this checklist.

Performance Assessment Checklist by Student

Diversity of Life

Investigation 2, Part 3

Student	Science and Engineering Practices		DCI	Crosscutting Concepts	
	Planning and carrying out investigations	Using mathematics and computational thinking	LS1.A Structure and function	Scale, proportion, and quantity	Structure and function

NOTE: See the Assessment chapter for a discussion about how to use this checklist.

Performance Assessment Checklist by Group

Diversity of Life

	Investigation 3, Part 1				
	Science and Engineering Practices		DCI	Crosscutting Concepts	
Group	Planning and carrying out investigations	Using mathematics and computational thinking	LS1.A Structure and function	Scale, proportion, and quantity	Structure and function

NOTE: See the Assessment chapter for a discussion about how to use this checklist.

FOSS Next Generation
© The Regents of the University of California
Can be duplicated for classroom or workshop use.

Performance Assessment Checklist by Student

Investigation 3, Part 1

Diversity of Life

Student	Science and Engineering Practices		DCI	Crosscutting Concepts
	Planning and carrying out investigations	Using mathematics and computational thinking	LS1.A Structure and function	Scale, proportion, and quantity	Structure and function

NOTE: See the Assessment chapter for a discussion about how to use this checklist.

FOSS Next Generation
© The Regents of the University of California
Can be duplicated for classroom or workshop use.

Diversity of Life Course
Performance Assessment Checklist
No. 7—Assessment Master

Performance Assessment Checklist by Group

Diversity of Life

Investigation 3, Part 3

Group	Science and Engineering Practices		DCI	Crosscutting Concept
	Planning and carrying out investigations	Using mathematics and computational thinking	LS1.A Structure and function	Scale, proportion, and quantity

NOTE: See the Assessment chapter for a discussion about how to use this checklist.

Performance Assessment Checklist by Student

Diversity of Life

Student	Investigation 3, Part 3			
	Science and Engineering Practices		DCI	Crosscutting Concept
	Planning and carrying out investigations	Using mathematics and computational thinking	LS1.A Structure and function	Scale, proportion, and quantity

FOSS Next Generation
© The Regents of the University of California
Can be duplicated for classroom or workshop use.

NOTE: See the Assessment chapter for a discussion about how to use this checklist.

Performance Assessment Checklist by Group

Diversity of Life

Investigation 5, Part 1

Group	Science and Engineering Practices					DCI	Crosscutting Concepts
	Planning and carrying out investigations	Analyzing and interpreting data	Using mathematics and computational thinking	Constructing explanations	Obtaining, evaluating, and communicating information	LS1.C Organization for matter and energy flow in organisms	Cause and effect

NOTE: See the Assessment chapter for a discussion about how to use this checklist.

FOSS Next Generation
© The Regents of the University of California
Can be duplicated for classroom or workshop use.

Diversity of Life Course
Performance Assessment Checklist
No. 10—Assessment Master

Performance Assessment Checklist by Student

Diversity of Life

Investigation 5, Part 1

Student	Science and Engineering Practices				DCI	Crosscutting Concepts
	Planning and carrying out investigations	Analyzing and interpreting data	Using mathematics and computational thinking	Constructing explanations	Obtaining, evaluating, and communicating information	LS1.C Organization for matter and energy flow in organisms	Cause and effect

NOTE: See the Assessment chapter for a discussion about how to use this checklist.

FOSS Next Generation
© The Regents of the University of California
Can be duplicated for classroom or workshop use.

Performance Assessment Checklist by Group

Diversity of Life

Investigation 6, Part 2

Group	Science and Engineering Practices			DCI	Crosscutting Concepts	
	Planning and carrying out investigations	Analyzing and interpreting data	Constructing explanations	LS1.B Growth and development of organisms	Patterns	Cause and effect

NOTE: See the Assessment chapter for a discussion about how to use this checklist.

Performance Assessment Checklist by Student

Diversity of Life

Investigation 6, Part 2

Student	Science and Engineering Practices			DCI	Crosscutting Concepts	
	Planning and carrying out investigations	Analyzing and interpreting data	Constructing explanations	LS1.B Growth and development of organisms	Patterns	Cause and effect

NOTE: See the Assessment chapter for a discussion about how to use this checklist.

Performance Assessment Checklist by Group

Diversity of Life

Investigation 8, Part 1

Group	Science and Engineering Practices			DCI	Crosscutting Concepts
	Asking questions	Planning and carrying out investigations	Analyzing and interpreting data	LS1.A Structure and function	Structure and function

NOTE: See the Assessment chapter for a discussion about how to use this checklist.

Performance Assessment Checklist by Student

Diversity of Life

Investigation 8, Part 1

Student	Science and Engineering Practices			DCI	Crosscutting Concepts
	Planning and carrying out investigations	Analyzing and interpreting data	Asking questions	LS1.A Structure and function	Structure and function

FOSS Next Generation
© The Regents of the University of California
Can be duplicated for classroom or workshop use.

NOTE: See the Assessment chapter for a discussion about how to use this checklist.

Performance Assessment Checklist by Group

Diversity of Life

| Group | Investigation 9, Part 1 ||||| Crosscutting Concepts |
| | Science and Engineering Practices |||| DCI | |
	Asking questions	Planning and carrying out investigations	Analyzing and interpreting data	Engaging in argument from evidence	LS2.C Ecosystem dynamics, functioning, and resilience	Systems and system models

NOTE: See the Assessment chapter for a discussion about how to use this checklist.

Performance Assessment Checklist by Student

Diversity of Life

Investigation 9, Part 1

Student	Science and Engineering Practices			DCI	Crosscutting Concepts
	Asking questions	Planning and carrying out investigations	Analyzing and interpreting data	Engaging in argument from evidence	LS2.C Ecosystem dynamics, functioning, and resilience	Systems and system models

NOTE: See the Assessment chapter for a discussion about how to use this checklist.

FOSS Next Generation
© The Regents of the University of California
Can be duplicated for classroom or workshop use.

Assessment Record—Entry-Level Survey

Diversity of Life

Item	Contributes to	Notes for Planning Instruction
1	MS-LS1-1	
2ab	MS-LS1-1 MS-LS1-3	
3	MS-LS1-1	
4	MS-LS1-3	
5	MS-LS1-5	
6	MS-LS1-4	
7	MS-LS3-2	

Assessment Record—Investigations 1–3 I-Check

Student names	1	2	3	4a	4b	5a	5b	6	7	8	9	10

Diversity of Life

FOSS Next Generation
© The Regents of the University of California
Can be duplicated for classroom or workshop use.

NOTE: A spreadsheet for this chart is available on FOSSweb.com

Assessment Record—Investigation 4 I-Check

Diversity of Life

Student names	1	2	3	4	5	6	7	8a	8b	8c	9	10

FOSS Next Generation
© The Regents of the University of California
Can be duplicated for classroom or workshop use.

NOTE: A spreadsheet for this chart is available on FOSSweb.com

Diversity of Life Course
Assessment Record
No. 20—Assessment Master

Assessment Record—Investigation 5 I-Check

Diversity of Life

Student names	1	2	3	4	5	6a	6b	7	8	9

FOSS Next Generation
© The Regents of the University of California
Can be duplicated for classroom or workshop use.

NOTE: A spreadsheet for this chart is available on FOSSweb.com

Assessment Record—Investigation 6 I-Check

Diversity of Life

Student names	1	2a	2b	3	4	5	6	7	8	9

NOTE: A spreadsheet for this chart is available on FOSSweb.com

Assessment Record—Investigation 7 I-Check

Diversity of Life

Student names	1	2	3	4	5	6	7a	7b	8

FOSS Next Generation
© The Regents of the University of California
Can be duplicated for classroom or workshop use.

NOTE: A spreadsheet for this chart is available on FOSSweb.com

Assessment Record—Posttest, 1 of 2

Diversity of Life

Student names	1	2	3	4a	4b	5	6	7	8	9

NOTE: A spreadsheet for this chart is available on FOSSweb.com

FOSS Next Generation
© The Regents of the University of California
Can be duplicated for classroom or workshop use.

Diversity of Life Course
Assessment Record

Assessment Record—Posttest, 2 of 2

Diversity of Life

Student names	10	11	12	13	14	15	16	17	18	19

NOTE: A spreadsheet for this chart is available on FOSSweb.com

ENTRY-LEVEL SURVEY
DIVERSITY OF LIFE

Name _____

Date _____ Class _____

1. A group of students are playing in a pile of dried oak leaves on the way home from school. They found a few acorns (oak tree seeds) lying near the pile of leaves.

    One student asked, "Are acorns and oak leaves alive?"

    Another student replied, "Acorns and leaves are not alive. They don't breathe, move, eat, or do any of the things living things do. How are they any different from a bunch of rocks?"

    Explain how you would respond to these students.

    _____
    _____
    _____
    _____
    _____
    _____
    _____
    _____

2. a. What are animals made of?

    _____
    _____
    _____
    _____

   b. What are plants made of?

    _____
    _____
    _____
    _____

ENTRY-LEVEL SURVEY
DIVERSITY OF LIFE

Name _____

3. List three ways that bacteria can affect your life.

   _____
   _____
   _____
   _____
   _____
   _____

4. Think about transport systems in animals and plants. In other words, how do all the parts of plants and animals get the resources they need to survive? How are the two transport systems alike and how are they different? Use the space at the bottom of the page to draw diagrams if that helps you answer this item.

   _____
   _____
   _____
   _____
   _____
   _____
   _____

ENTRY-LEVEL SURVEY
DIVERSITY OF LIFE

Name _____

5. Seeds for two different kinds of maple trees were planted in a yard. One grew into a tall and healthy tree. The other did not grow as tall and was weak.

   What factors may have influenced the growth of the two different kinds of trees?

   _____
   _____
   _____
   _____
   _____
   _____
   _____

6. Describe how pollinators (like insects, birds, and bats) and flowering plants depend on each other for survival and reproduction.

   _____
   _____
   _____
   _____
   _____
   _____
   _____
   _____

ENTRY-LEVEL SURVEY
DIVERSITY OF LIFE

Name _____

7. Two dogs with brown fur mate and have a litter of puppies. In the litter, two puppies have golden fur, two puppies have brown fur, and one puppy has black fur.

   Explain how this is possible.

INVESTIGATIONS 1–3 I-CHECK
DIVERSITY OF LIFE

Name _____

Date _____ Class _____

1. Write **L** next to each phrase that is a characteristic of all living systems; write **N** if the phrase does not describe a characteristic of all living systems.

   _____ Reproduces   _____ Needs water

   _____ Exchanges gas   _____ Has a pulse

   _____ Moves   _____ Responds to the environment

2. Write **E** next to each class investigation that provided evidence that living things are made of one or many cells; write **N** if the statement does not provide that evidence.

   _____ Rubbing the inside of your cheek and using a microscope to view the wet mount

   _____ Learning the parts of a microscope and how to use it

   _____ Observing sand, yeast, polyacrylate beads, radish seeds, and brine shrimp eggs

   _____ Looking at elodea leaves through a microscope

   _____ Observing paramecia with the aid of a microscope

   _____ Determining how field of view changes when you use different lenses

3. Two students were looking at a leaf that was brown and dried out. Student A said, "I would call this nonliving." Student B said, "I think it's better to say that it is dead."

   Which student do you agree with and why?

   (Mark the one best answer.)

   ○ **A** Student A because a leaf is not a living part of a plant

   ○ **B** Student A because a leaf can't breathe or reproduce

   ○ **C** Student B because a leaf was once living

   ○ **D** Student B because a leaf is made of cells

INVESTIGATIONS 1–3 I-CHECK
DIVERSITY OF LIFE

Name _____

4. A seventh grade student was walking home with his 5-year-old brother. The younger boy said, "I think clouds must be alive. They move across the sky, they start small and get bigger, they need air, and they need water to grow."

   a. Why might a young child think a cloud was alive?

   _____

   _____

   _____

   b. What would you tell the younger boy to convince him that clouds are not alive?

   _____

   _____

   _____

   _____

5. A student used a microscope to look at an amoeba and a millimeter ruler. First she used 40X magnification, and then she turned the objective lens and used 100X magnification.

   a. At 40X magnification, how wide is the field of view? _____

   How big is the amoeba at 40X? _____

   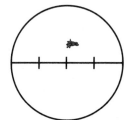

   b. At 100X magnification, how wide is the field of view? _____

   Is the size of the actual amoeba on this slide bigger, smaller, or the same size as the one shown above it? _____

INVESTIGATIONS 1–3 I-CHECK
DIVERSITY OF LIFE

Name _____

6. Explain the difference between a single cell that is considered an organism and a single cell that could not be considered an organism.

_____
_____
_____
_____
_____
_____
_____

7. The cell in the image has green dots moving in it. What kind of cell do you think it is and why?

    green dots

    (Mark the one best answer.)

    ○ **A** Plant cell because the green dots are chloroplasts

    ○ **B** Plant cell because it has a cell membrane

    ○ **C** Animal cell because the green dots are probably the stained organelles

    ○ **D** Animal cell because the cell walls are rigid

8. Science often depends on engineering advances. Which tool allowed scientists to observe organisms too small to see with the naked eye?

    (Mark the one best answer.)

    ○ **A** Telescope

    ○ **B** Stethoscope

    ○ **C** Spectroscope

    ○ **D** Microscope

INVESTIGATIONS 1–3 I-CHECK
DIVERSITY OF LIFE

Name _____

9. Use the word bank to fill in the names of the labeled cell structures. Write the function of each cell structure next to its name.

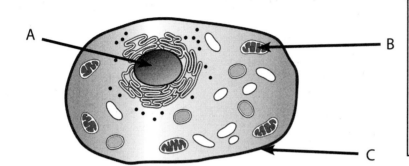

Word Bank
cell membrane
cell wall
lysosome
mitochondrion
nucleus
ribosome

	Name	Function
A		
B		
C		

10. Which statement describes what happens during asexual reproduction of single-celled organisms like paramecia?

(Mark the one best answer.)

○ **A** The mother cell divides and is replaced by two or three daughter cells.

○ **B** All of the cell structures of the mother cell divide in half during reproduction.

○ **C** The daughter cells have the same genetic material as the mother cell.

○ **D** After the mother cell splits in two, each daughter cell is exactly the same size and shape as the mother cell.

INVESTIGATION 4 I-CHECK
DIVERSITY OF LIFE

Name _____

Date _____ Class _____

1. Write **M** in the blank if the organism has a cell *membrane* only; write **W** if the organism has a cell *wall* in addition to a cell membrane.

    _____ Paramecium   _____ Fungus

    _____ Bacterium   _____ Human cheek cell

    _____ Archaea   _____ Elodea cell

2. Bacteria reproduce asexually—they simply duplicate themselves. Explain how bacteria acquire new characteristics if this is true.

    _____
    _____
    _____
    _____

3. Which organisms, archaea, bacteria, or fungi, are most like tomato plants? What is your reasoning for choosing that answer?

    _____
    _____
    _____
    _____
    _____
    _____
    _____
    _____

INVESTIGATION 4 I-CHECK
DIVERSITY OF LIFE

Name _____

4. Write **E** next to each statement that helps explain why scientists didn't realize until the 1970s that archaea are a unique kind of life, different from bacteria. Write **N** if the statement does not help explain why that was the case.

   _____ Archaea are a lot like bacteria.

   _____ Both archaea and bacteria have cell walls.

   _____ Before the 1970s microscopes weren't powerful enough to tell the difference.

   _____ Archaea are found only in extreme environments.

5. Yeast are used in making bread. How do yeast cause bread to rise?

   _____
   _____
   _____
   _____
   _____

6. Write **P** next to each statement that describes how most people think of bacteria interacting *positively* in the world; write **N** next to each statement that describes how most people think of bacteria interacting *negatively* or *neutrally*.

   _____ The decomposition of dead organisms

   _____ Food digestion in the guts of multicellular organisms

   _____ Food production such as cheese and yogurt

   _____ Making people sick

   _____ Killing other bacteria

INVESTIGATION 4 I-CHECK
DIVERSITY OF LIFE

Name _____

7. Observe the drawing below.

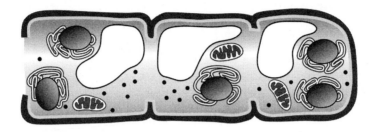

Write **U** next to each statement that indicates a *unique* characteristic of fungus; write **N** next to each statement that is not a unique characteristic of fungus.

_____   There are three cells grouped together.

_____   These cells have mitochondria.

_____   There are holes in the cell walls.

_____   Some of the cells have more than one nucleus.

8. a. Name the three domains:

   _____

   _____

   _____

   b. Which domain includes multicellular organisms? _____

   c. How else is the domain you named in item b different than the other two domains?

   _____
   _____
   _____
   _____
   _____
   _____

INVESTIGATION 4 I-CHECK
DIVERSITY OF LIFE

Name _____

9. Bacteria and archaea are similar because both _____.

   ○ **A** are eukaryotes

   ○ **B** are prokaryotes

   ○ **C** have the same kind of cell wall

   ○ **D** store their DNA in a nucleus

10. Use the words from the word bank to list the levels of complexity in order from atoms to organisms. (You will not use all the words.)

    atom

    _____

    _____

    _____

    _____

    _____

    multicellular organism

    Word bank
    archaea
    bacteria
    cell
    cell structure/organelle
    molecule
    organ
    organ system
    tissue

INVESTIGATION 5 I-CHECK
DIVERSITY OF LIFE

Name _____

Date _____ Class _____

1. Describe the process of transpiration, from the time water enters a plant through the roots to when water leaves a plant.

   _____
   _____
   _____
   _____
   _____
   _____
   _____
   _____
   _____
   _____
   _____
   _____
   _____
   _____

2. How is transpiration related to photosynthesis?
   (Mark the one best answer.)

   ○ **A** They are not related; they are two separate processes that do not interact.

   ○ **B** Transpiration brings water to cells; photosynthesis carries it away.

   ○ **C** Transpiration delivers water needed for photosynthesis to the cells.

   ○ **D** Photosynthesis creates water and sugar in a chemical reaction and transpiration delivers the sugar dissolved in water to the plant's root cells for food.

INVESTIGATION 5 I-CHECK
DIVERSITY OF LIFE

Name _____

3. If you place a plastic bag over a branch of a tree that has leaves during a drought (a time with very little or no rain), how much moisture would you expect to see in the bag a few hours later? Explain why that will happen.

_____
_____
_____
_____

4. Which of the following statements describes where the processes of photosynthesis and aerobic cellular respiration occur?
   (Mark the one best answer.)

   - ○ A Photosynthesis and cellular respiration occur only in plants.
   - ○ B Photosynthesis occurs only in plants, and cellular respiration occurs only in animals.
   - ○ C Photosynthesis and cellular respiration occur in both plants and animals.
   - ○ D Photosynthesis occurs only in plants, and cellular respiration occurs in both plants and animals.

5. How do plant cells that do not have chlorophyll get food?
   (Mark the one best answer.)

   - ○ A Phloem carries food from the leaves to all the cells in a plant.
   - ○ B Xylem carries food from the leaves to all the cells in a plant.
   - ○ C All plant cells have chlorophyll, so there is no problem.
   - ○ D The plant cells that do not have chlorophyll do not need food.

INVESTIGATION 5 I-CHECK
DIVERSITY OF LIFE

6. Student A left a piece of celery in a vial overnight. The next morning he noticed that the water level had decreased.

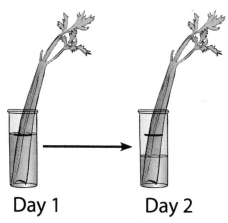

Day 1   Day 2

Student B walked by and said, "Wow, a lot of water evaporated from that vial." Student A said, "I don't think much evaporated. I think most of the change in water level was due to transpiration."

a. Write an argument that would support student A's conclusion.

___

b. Write an argument that would support student B's conclusion.

___

INVESTIGATION 5 I-CHECK
DIVERSITY OF LIFE

7. The stems and leaves in a plant are examples of plant _____.

    (Mark the one best answer.)

    ○ **A** organs

    ○ **B** organ systems

    ○ **C** molecules

    ○ **D** tissues

8. Photosynthesis uses energy from the Sun to make sugars such as glucose. What has to happen to the glucose before it can be used by plant cells?

    (Mark the one best answer.)

    ○ **A** The glucose combines with another sugar to make usable energy.

    ○ **B** The glucose combines with oxygen to make usable energy.

    ○ **C** The glucose combines with carbon dioxide to make usable energy.

    ○ **D** Nothing has to happen to the glucose. Plant cells can use it as is.

9. The chemical equation for photosynthesis is

    $$6\,CO_2 + 6\,H_2O + \text{light energy} \longrightarrow C_6H_{12}O_6 + 6\,O_2$$

    Write the chemical equation for aerobic cellular respiration.

    _____

INVESTIGATION 6 I-CHECK
DIVERSITY OF LIFE

Name _____

Date _____ Class _____

1. Write **L** next to each statement that is evidence that seeds are *living* things; write **N** next to each statement that does not provide evidence or is not true.

    _____ A seed makes its own food using photosynthesis.

    _____ Seeds are the product of sexual reproduction.

    _____ Seeds come in many different shapes and sizes.

    _____ Seeds are food for animals.

    _____ A seed has an embryo.

    _____ Seeds are dormant until placed in a suitable environment.

2. a. A hummingbird (seen at right) is most likely to pollinate which of the following flowers?
    (Mark the one best answer.)

    ○ A

    ○ B

    ○ C

    ○ D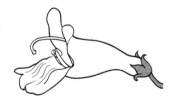

    b. Explain why you chose that flower.

    _____

    _____

    _____

INVESTIGATION 6 I-CHECK
DIVERSITY OF LIFE

3. Use the word bank to list the steps in plant reproduction.

   (1) _____

   (2) _____

   (3) _____

   (4) _____

   Word bank
   fertilization
   pollen-tube formation
   pollination
   seed formation

4. A group of university researchers recently reported that beekeepers lost 42.1 percent of their colonies in 2015. This is the second-highest loss recorded since year-round surveys began in 2010. Why are farmers and ecologists concerned by the loss of these honey bees?

   _____

   _____

   _____

   _____

   _____

   _____

   _____

   _____

   _____

5. Increasing the salinity of soil can _____.
   (Mark the one best answer.)

   ○ A cause the phloem of a plant to transport more food to the leaves

   ○ B cause plant leaves to transpire more

   ○ C affect a plant's genetic factors

   ○ D keep a plant's seeds from germinating

INVESTIGATION 6 I-CHECK
DIVERSITY OF LIFE

Name _____

6. The _____ is the primary source of energy for seedlings in the early days of growth.

   (Mark the one best answer.)

   ○ **A** Sun

   ○ **B** cotyledon

   ○ **C** embryo

   ○ **D** shoot

7. Frost tolerance is a plant's ability to survive below-freezing temperatures. Summarize the steps you would take to investigate the frost tolerance of petunias (a flowering plant). Be sure to include the evidence you would look for to determine if the petunias are frost tolerant.

   _____
   _____
   _____
   _____
   _____
   _____
   _____
   _____
   _____
   _____
   _____
   _____
   _____
   _____
   _____
   _____

INVESTIGATION 6 I-CHECK
DIVERSITY OF LIFE

8. Fish that live in the ocean have characteristics that allow them to live in salt water. This is probably due to the fish's _____.

   (Mark the one best answer.)

   ○ **A** fertilization factors

   ○ **B** growth factors

   ○ **C** environmental factors

   ○ **D** genetic factors

9. Write the function of each of the image's labeled structures on the lines below.

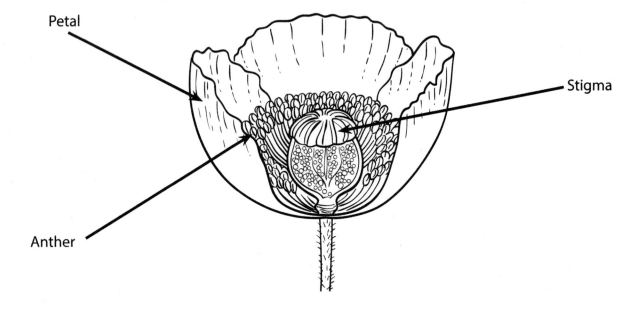

Anther _____

_____

Petal _____

_____

Stigma _____

_____

INVESTIGATION 7 I-CHECK
DIVERSITY OF LIFE

Name _____

Date _____ Class _____

1. Match the definitions and words. Write the letter that represents each word in front of its definition.

    WORDS
    (in alphabetical order)

    C chromosome
    D dominant
    G genotype
    H heredity
    P phenotype
    R recessive
    T trait
    V variation

    _____ Explains why organisms are similar but not identical to their parents.

    _____ The individual expression of a feature.

    _____ The genes that make up an organism.

    _____ The structures made of DNA that transfer genetic information to the next generation.

    _____ An allele that is expressed in offspring only when inherited from both parents.

    _____ The traits that you can observe in an organism.

    _____ The range of difference between the traits of a feature in the individuals of a population.

2. Write **T** in the blank if the statement is true; write **F** in the blank if the statement is false.

    _____ Some organisms do not have DNA.

    _____ An allele is a version of a gene.

    _____ Cells in a dog's stomach contain the same DNA as cells in the dog's eyes.

    _____ Sex cells (egg and sperm cells) contain half the number of chromosomes as cells in other parts of the organism.

    _____ Under normal circumstances, offspring have twice as many chromosomes (and therefore genes) in their cells as their parents.

INVESTIGATION 7 I-CHECK
DIVERSITY OF LIFE

3. A student's aunt is about to have a baby. The aunt has freckles (she has two dominant genes for the freckles trait) and the aunt's husband does not have freckles (he has two recessive genes for the freckles trait). Assume the freckles trait follows a simple dominant/recessive inheritance pattern.

   Complete the Punnett square to determine the probability of the baby having freckles.

   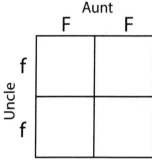

   What is the probability that their child will have freckles? _____

4. Sea anemones can reproduce sexually or asexually. When they reproduce asexually, they go through a process called budding. How much of the DNA in the offspring sea anemone created by budding comes from the parent sea anemone?

   (Mark the one best answer.)

   ○ **A** All of the DNA

   ○ **B** Half of the DNA

   ○ **C** None of the DNA

   ○ **D** The amount can't be predicted in this situation.

INVESTIGATION 7 I-CHECK
DIVERSITY OF LIFE

5. Here is a list of plants and animals and the number of chromosomes in their cells.

Potato	48	Mango	40	Horse	64	Dolphin	44
Bean	22	Black mulberry	308	Fruit fly	8	Human	46
Pea	14	Pineapple	50	Bear	74	Snail	24

What pattern do you notice?

_____

_____

Why do you think that pattern occurs? Explain the mathematics and give an example.

_____

_____

_____

_____

_____

6. A student noticed that she looks more like her mother's side of the family than her father's. She explained, "That's because girls inherit DNA from their mothers and boys inherit DNA from their fathers."

Explain why you agree or disagree with this student's statement.

_____

_____

_____

_____

_____

_____

Name _____

INVESTIGATION 7 I-CHECK
DIVERSITY OF LIFE

7. When a person has a genetic hearing loss, he or she is born deaf. The most common genetic hearing loss is caused by a homozygous recessive genotype.

   A man and a woman want to marry and start a family. The woman has good hearing and has no history of hearing loss in her family. The man has good hearing as well, but his mother has a genetic hearing loss.

   a. What is the probability that their first born child will have a genetic hearing loss? (Create and show the model you use to determine the probability.)

   _____

   Work space

   b. What is the probability that their second born will have a genetic hearing loss?

   _____

8. Read each statement. Write **S** in the blank if the statement describes only sexual reproduction. Write **A** if the statement describes only asexual reproduction. Write **B** if the statement describes both.

   _____ Offspring are produced.

   _____ Each offspring receives genetic information from two parents.

   _____ Egg and sperm cells fuse.

   _____ The genetic information in the offspring is exactly like that of the parent.

   _____ Results in offspring with genetic variation.

POSTTEST
DIVERSITY OF LIFE

Name _____

Date _____ Class _____

1. Biodiversity is a measure of _____.
   (Mark the one best answer.)

   ○ **A** the variety of species and number of organisms in an area

   ○ **B** the diversity of environmental factors in an area

   ○ **C** the quality of organisms in a certain area

   ○ **D** the quality of life in a habitat

2. Explain why a cell is considered the basic unit of life.

   _____
   _____
   _____
   _____

3. In the spring of each year, male bowerbirds in Australia build elaborate nests using sticks and brightly colored objects. Describe how this behavior might contribute to successful reproduction.

   _____
   _____
   _____
   _____
   _____
   _____
   _____
   _____

POSTTEST
DIVERSITY OF LIFE

Name _____

4. Use the image below to answer items a and b.

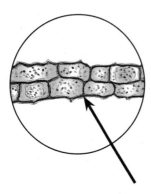

a. The cells in this image of the microscope's field of view are most likely from a(n) _____.
   (Mark the one best answer.)

   ○ **A** bacterium   ○ **B** plant

   ○ **C** protist   ○ **D** fungus

b. The arrow is pointing to a cell structure. The function of the structure is to _____.
   (Mark the one best answer.)

   ○ **F** turn food into usable energy

   ○ **G** contain the cell's DNA

   ○ **H** regulate the production of proteins

   ○ **J** provide rigid structure and protection for the cell

5. A student saw something moving on the floor and exclaimed, "It's alive!" Write an argument that includes evidence and explains why you agree or disagree with this student.

   _____
   _____
   _____
   _____
   _____
   _____

POSTTEST
DIVERSITY OF LIFE

Name _____

6. Aerobic cellular respiration _____.

   (Mark the one best answer.)

   ○ **A** gets rid of waste made by cells

   ○ **B** happens mostly in muscle cells

   ○ **C** turns food into glucose molecules in cells

   ○ **D** turns glucose into usable energy in cells

7. The purpose of photosynthesis is to make _____.

   (Mark the one best answer.)

   ○ **A** sugar(s)

   ○ **B** new leaves for a plant

   ○ **C** usable energy for animals

   ○ **D** light

8. All organisms carry out all the functions of life. Explain the difference in how multicellular and single-celled organisms do this.

   _____
   _____
   _____
   _____
   _____
   _____
   _____
   _____
   _____

POSTTEST
DIVERSITY OF LIFE

Name _____

9. Consider flowering plant reproduction. Write **T** if the sentence is true; write **F** if the sentence is false.

   _____ Flowers are the reproductive organs of plants.

   _____ Flowers attract pollinators to make sexual reproduction possible.

   _____ Pollen is the carrier of the egg cell.

   _____ During the process of pollination, pollen is transported to the stigma and sperm travels down a pollen tube to the egg.

   _____ Fertilized eggs develop into seeds, which hold the potential for a new plant.

10. Most seeds germinate and start to grow in the dark. Why don't they need light right away? *(Mark the one best answer.)*

    ○ **A** The embryo stores food the plant needs to start growing.

    ○ **B** The cotyledon stores food the plant needs to start growing.

    ○ **C** The roots store food the plant needs to start growing.

    ○ **D** Seeds photosynthesize to make the food the plant needs to start growing.

11. An elodea cell is not considered an organism, but a paramecium cell is. Explain why this is true.

    _____
    _____
    _____
    _____
    _____
    _____
    _____

POSTTEST
DIVERSITY OF LIFE

Name _____

12. The function of xylem in a plant is to _____.
    (Mark the one best answer.)

    ○ **A** perform photosynthesis

    ○ **B** transport food to all cells of the plant

    ○ **C** transport water to all cells of the plant

    ○ **D** give rigid structure to plant cells

13. Which of the following statements is true of aerobic cellular respiration?
    (Mark the one best answer.)

    ○ **A** All living things do it.

    ○ **B** Only plants do it.

    ○ **C** Only animals do it.

    ○ **D** All living things except plants do it.

14. Bacteria usually reproduce using _____.
    (Mark the one best answer.)

    ○ **A** asexual reproduction

    ○ **B** sexual reproduction

    ○ **C** hyphae and plasmids

    ○ **D** atypical reproduction

POSTTEST
DIVERSITY OF LIFE

Name _____

15. A student found a little pile of material that looked like tiny beads near a tree on her street. She thought that they were dormant organisms. What did she mean by that?

_____
_____
_____
_____
_____
_____
_____
_____
_____
_____

16. What is the most significant difference between bacteria and eukaryotes?
    (Mark the one best answer.)

    ○ **A** Eukaryotic cells don't have a cell wall.

    ○ **B** Eukaryotes are always multicellular organisms.

    ○ **C** Eukaryotic cells have a nucleus.

    ○ **D** Eukaryotic cells don't have a nucleus.

Name _____

POSTTEST
DIVERSITY OF LIFE

17. Use words from the word bank to list the levels of complexity in order from atom to organism.

    atom

    _____

    _____

    _____

    _____

    _____

    _____

    multicellular organism

Word bank
cell
cell structure/organelle
molecule
organ
organ system
tissue

18. Wheat is being bred to be more salt tolerant. This means that scientists are experimenting with the wheat's _____.
    (Mark the one best answer.)

    ○ **A**   germination factors

    ○ **B**   environmental factors

    ○ **C**   genetic factors

    ○ **D**   planting factors

19. Why is photosynthesis important to animals?
    (Mark the one best answer.)

    ○ **A**   Animals make their own food with the process of photosynthesis.

    ○ **B**   Plants use photosynthesis to make food that animals can eat.

    ○ **C**   Bacteria use photosynthesis to help animals digest food.

    ○ **D**   Animals don't use photosynthesis, so it is not important to them.

Notebook Answers

Living/Nonliving Card Sort

Card name	L	NL	U	Card name	L	NL	U
Amoeba				Mushrooms			
Apple				Onions			
Baby				Potatoes			
Blue cheese				Rhinovirus			
Blue-green algae				Robot			
Bread mold				Rocking horse			
Cactus				Spider and web			
Clouds							
Coral							
Corn							
Cotton boll							
E. coli				Yeast			
Eggs				Yogurt			
Fire							
Horse							
Jellyfish							
Kelp							

Answers will vary. Do not attempt to correct.

Five-Materials Observation

Liquid number _____

	A	B	C	D	E
First observations (dry) (include drawings)					
Changes observed after 10 minutes (include drawings)					
Changes observed after 24 hours (include drawings)					
Changes observed after _____ (include drawings)					

Answers will vary. Look for drawings and evidence of change where appropriate.

The Microscope

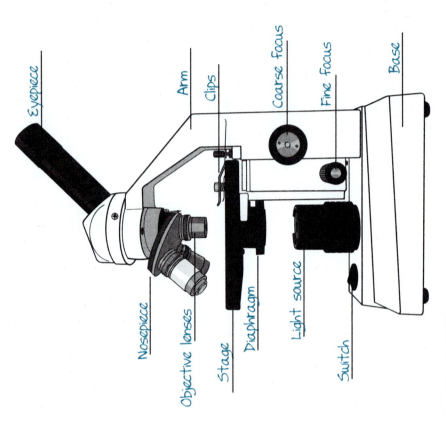

Microscope Care

- Always use two hands to carry a microscope—one hand holding the **arm** and one hand under the **base**. If the microscope has a power cord, gather that up to keep it from getting underfoot.
- Always wipe up any water that is on the scope after a session.
- Always cover the microscope with a dustcover to keep it clean.
- Use *only* lens paper to clean the lenses of a microscope. Do not use paper towels or tissues as they will scratch the lenses.

Life in Different Environments

Liquid 1 Salty water

	Material	What evidence of life do you observe?
A	Red sand	None
B	Yeast	None
C	Polyacrylate	None
D	Radish seeds	Nothing, or seeds started to open
E	Brine shrimp eggs	Brine shrimp hatched

Liquid 2 Sugar water

	Material	What evidence of life do you observe?
A	Red sand	None
B	Yeast	Bubbles; cap popped; smells like bread
C	Polyacrylate	None
D	Radish seeds	Sprouting
E	Brine shrimp eggs	None

Liquid 3 Plain water

	Material	What evidence of life do you observe?
A	Red sand	None
B	Yeast	None
C	Polyacrylate	None
D	Radish seeds	Sprouting
E	Brine shrimp eggs	None

Microscope Images

Part 1: Draw the letter e under low power.

1. Set the objective lens to 4X.
2. Place the slide of the word *seed* on the stage of the microscope *so you can read it*.
3. Center the image in the field of view on one *e* and draw *exactly* what you see.

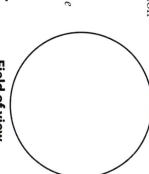

Field of view

Part 2: Move the slide.

4. Move the slide away from you. What direction did the image move? __Toward me__
5. Move the slide to your right. What direction did the image move? __Left__

Part 3: Redraw the letter e after discussion.

6. *After the class discussion, redraw the letter e here.*

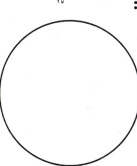

Field of view

Drawings will vary. The "e" should be upside down and backward (depending upon your microscopes) and show detail.

Part 4: Answer these questions in your science notebook.

7. Is the image seen through the microscope oriented the same way as the object on the stage of the microscope? Explain.
8. If you want to move the image to the right, which way should you move the slide?
9. If you want to move the image up, which way should you move the slide?

7. No. The image is upside down and backward.
8. To the left
9. Toward me

Field of View and Magnification

Part 1: Use the 4X objective.
1. At low power, what is the width of the field of view? __~4 mm__
2. What is the total magnification with this objective lens? __40X__

Part 2: Use the 10X objective.
3. Place the slide on the stage and place the clear millimeter ruler on top, using the frame of reference your class decided upon.
4. Draw exactly what you see.

   At medium power, what is the width of the field of view? __~1.5 mm__

   What is the total magnification? __100X__

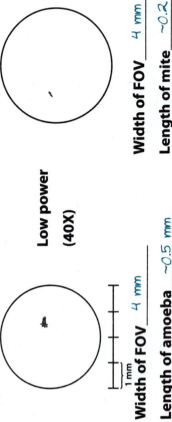

Part 3: Use the 40X objective.
5. Change to high power.
6. Draw exactly what you see.

   At high power, what is the width of the field of view? __~0.4 mm__

   What is the total magnification? __400X__

Part 4: Mark the scale.
7. In part 2, for the 10X objective, mark the scale in millimeters on the line under the field of view.
8. In part 3, for the 40X objective, mark the scale in *tenths* of millimeters on the line under the field of view. (Careful!)

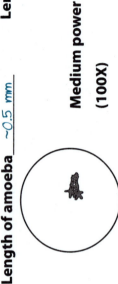

Part 5: How big is the letter e?
9. Refer to your notebook and estimate the width of the letter *e* you drew earlier under low power. __2 mm__
- How wide would the *e* appear under medium power? __>1.5 mm__
- How wide would the *e* appear under high power? __would be hard to tell__

Estimating Size

A student determined the diameter of the field of view (FOV) of his microscope at each magnification. He drew an amoeba at each magnification and then drew a follicle mite at each magnification. Estimate the diameter of the FOV and how long each organism is at each magnification.

Low power (40X)

Width of FOV __4 mm__   Length of mite __~0.2__

Width of FOV __4 mm__   Length of amoeba __~0.5 mm__

Medium power (100X)

Width of FOV __2 mm__   Length of mite __~0.2__

Width of FOV __2 mm__   Length of amoeba __~0.5 mm__

High power (400X)

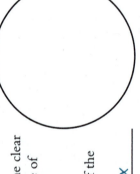 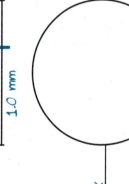

Width of FOV __0.4 mm__   Length of mite __~0.2__

Width of FOV __0.4 mm__   ~0.5 mm   Length of amoeba __~0.5 mm__

Response Sheet—Investigation 2

A student told her friend,

> We looked at things through the microscope today. I saw something called a mite. It got bigger and bigger and bigger as I increased the magnification.

The student's friend looked confused and said,

> I am not sure what you mean by "bigger and bigger."

How should the first student correctly explain herself?

The size of the mite only appeared to get bigger and bigger as I increased the magnification. The actual size of the mite did not change.

Brine Shrimp

WARNING—This set contains chemicals that may be harmful if misused. Read cautions on individual containers carefully. Not to be used by children except under adult supervision.

Part 1: Observe brine shrimp in the vial.

1. How do brine shrimp respond to light? See your teacher's demonstration or shine a flashlight through the vial.

   The brine shrimp swim toward the light.

2. Compare the size of the brine shrimp now to the size of the brine shrimp when they first hatched. How are they different?

   The brine shrimp are bigger now. When they first hatched, I had to use a hand lens to see them.

Part 2: Observe brine shrimp under the microscope.

3. Use a dropper to take up a few brine shrimp. Put one drop on the surface of a slide. If no brine shrimp are on the slide, wipe the slide dry and put on another drop.
4. Use a piece of blotter paper to soak up part of the water.
5. Do not put a coverslip on the slide.
6. Observe and draw an illustration of the brine shrimp at **100X**.
7. Estimate the size of the brine shrimp.

   ≈2–8 mm

Medium power (100X)

Part 3: Add yeast to the brine shrimp.

8. Carefully add one tiny drop of Congo red–dyed yeast to the slide.
9. Observe the yeast and the brine shrimp. Describe what you see.

   The brine shrimp are eating the yeast. You can see the red yeast inside the shrimp's body.

Plant Cell Structures and Functions

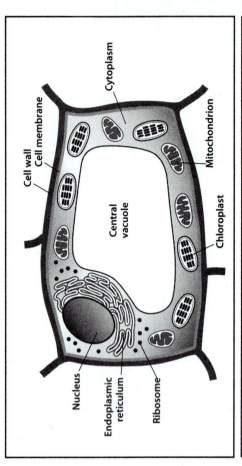

Cell structure	Function
Cell membrane	Cell boundary that controls what enters and leaves the cell
Cell wall	A rigid layer that supports the cell and provides shape
Central vacuole	Stores water and other substances; provides structure and support for the plant cell
Chloroplast	Converts the Sun's energy into food (sugars)
Cytoplasm	Internal fluid that contains the cell structures
Endoplasmic reticulum	A membranous structure that assembles proteins and parts of the cell membrane
Mitochondrion	Converts the energy in food into usable energy for the cell
Nucleus	Contains the cell's genetic material (DNA), which determines the nature of cell structures and substances
Ribosome	A structure that makes proteins either free or bound to the surface of the endoplasmic reticulum

Looking at Elodea

Part 1: Observe elodea leaf layers.

1. Place a small elodea leaf on a slide, top side up, bottom side against the slide. Prepare a wet mount using pond water and a coverslip.
2. Focus the microscope at 40X and then increase to 100X.
3. Increase the magnification to 400X. Using the fine focus knob, carefully focus up and down through the different layers of the leaf. How many layers can you see? __2__
4. Describe what you observe.

   Green rectangles with little green dots or circles inside. The green rectangles seem to be different sizes in the different layers.

Part 2: Observe elodea details and cell size.

5. Look carefully for movement inside the leaf. Describe what you observe.

   The green dots are moving around the edges of the rectangle.

6. Draw a few representative *large* brick-like structures to scale in the circle. Do not fill in the entire field of view. Use color and include detail.

High power (400X)

7. How many of the *large* green "bricks" fit lengthwise across the field of view? __3 or 4__
8. Estimate the size of one of the "bricks." __0.1–0.15 mm__

Part 3: Label the drawing.

9. Label the cell wall, chloroplasts, and cytoplasm.

WARNING — This set contains chemicals that may be harmful if misused. Read cautions on individual containers carefully. Not to be used by children except under adult supervision.

Paramecia

Part 1: Observe movement and behavior.

1. Put one small drop of paramecium culture on the center of your slide. Do NOT put on a coverslip.
2. Focus the microscope at 40X to make sure you have paramecia on your slide. Increase the magnification to 100X.
3. Describe the movement and behavior of the paramecia.

   moving through the water, spinning, turning, rolling over, bumping into things

Part 2: Observe paramecium up close.

4. Remove the slide from the stage and add one drop of methyl cellulose. Put on a coverslip. If necessary, blot up extra liquid.
5. Find one paramecium that is still moving, focus under low power, and increase to medium and then to high power. Focus using the fine focus knob. Describe the paramecium and draw it in the circle below. *The paramecium is a long oval with a fold (groove) on one end. There are little circles inside the paramecium and little hairs (legs) all around the outside.*
6. Estimate the length of the paramecium. _<0.5 mm_

Part 3: Label the drawing.

7. Label the cell membrane, cytoplasm, cilia, and any other structures you observe.
8. What is the purpose of the cell membrane? *The cell membrane holds all the stuff inside the paramecium, lets some things in and out, and protects it.*

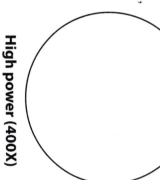

High power (400X)

Protist Cell Structures and Functions

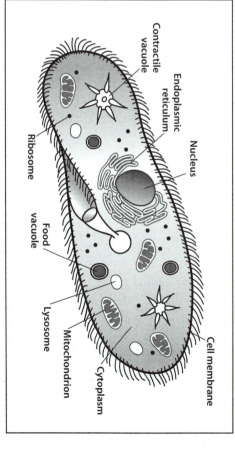

Cell structure	Function
Cell membrane	Boundary that controls what enters and leaves the cell
Contractile vacuole	A membrane that stores water and expels excess water
Cytoplasm	Internal fluid that contains the cell structures
Endoplasmic reticulum	A membranous structure that assembles proteins and parts of the cell membrane
Food vacuole	A membrane that stores food and merges with a lysosome to digest food
Lysosome	An organelle that digests cellular waste and merges with a food vacuole to digest food
Mitochondrion	An organelle that converts the energy in food into usable energy for the cell
Nucleus	An organelle that contains the cell's genetic material (DNA), which determines the nature of cell structures and substances
Ribosome	A structure that makes proteins, either free or bound to the endoplasmic reticulum

Minihabitat Safari

Is there anything living in the minihabitat?

1. Prepare a wet mount from one region of your minihabitat. Look for life at 40X.
2. If necessary, add one drop of methyl cellulose. Put on a coverslip and blot away any extra liquid. Increase the magnification to 100X and then 400X as needed.
3. Draw to scale any organisms you observe. Use the next page in your science notebook to describe their behavior and to add more organisms.
4. Use "Microorganism Guide" in *Science Resources* to help identify any organisms you find.

WARNING — This set contains chemicals that may be harmful if misused. Read cautions on individual containers carefully. Not to be used by children except under adult supervision.

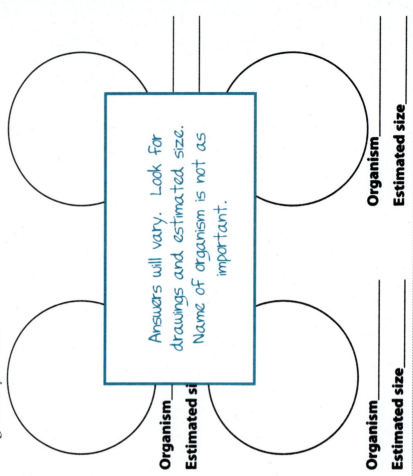

Answers will vary. Look for drawings and estimated size. Name of organism is not as important.

Organism _____
Estimated size _____

Organism _____
Estimated size _____

Organism _____
Estimated size _____

Organism _____
Estimated size _____

Response Sheet—Investigation 3

Two students were having a discussion. One said,

All cells are living things. Every cell in an elodea plant is an organism, just like the one-celled paramecium we looked at.

The second student said,

Well, you're partly right. I agree that all cells are living things, but an elodea cell is not an organism.

Evaluate what each student said. Explain your thinking.

First student:

Yes, all cells are living, but an elodea cell is not an organism itself. It is one cell in a multicellular organism. It cannot live on its own. An elodea plant is an organism, but a single elodea cell is not.

Second student:

This student is correct. All cells are living things, but a single elodea cell is not an organism. It is part of the larger multicellular organism, an elodea plant.

Human Cheek Tissue

WARNING — This set contains chemicals that may be harmful if misused. Read cautions on individual containers carefully. Not to be used by children except under adult supervision.

Part 1: Prepare a cheek-tissue sample.

1. Gently rub the inside of your cheek with a cotton swab.
2. Roll the rubbing onto the center of a slide. Add one drop of methylene blue and let set for 1 minute.
3. Hold the slide over a waste container and rinse it with a few drops of water. Add a drop of water if necessary, place a coverslip on top, and blot any extra water from the edges.
4. View the slide, starting at 40X. Use the search image your teacher provides to help you focus on the stained cheek tissue. Increase magnification to 400X.

Part 2: Record observations.

5. Describe what you see at 400X and draw it in the circle.

   Small, irregularly shaped cells with a dark dot in the middle

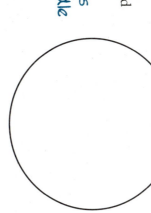

High power (400X)

6. Estimate the diameter of one cell.

   0.03–0.05 mm

Part 3: Questions

7. What is the inside of your cheek made of?

   Cells

8. What do you think other parts of your body are made of?

   Different kinds of cells for different parts

9. Label the cell membrane and nucleus in one of the cheek cells.

Clean up as directed.

Animal Cell Structures and Functions

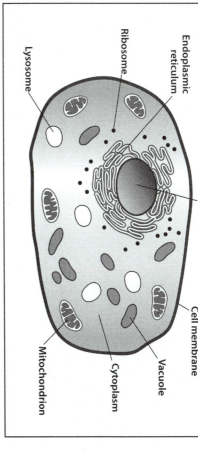

Cell structure	Function
Cell membrane	Boundary that controls what enters and leaves the cell
Cytoplasm	Internal fluid that contains the cell structures
Endoplasmic reticulum	A membranous structure that assembles proteins and parts of the cell membrane
Lysosome	An organelle that digests cellular waste
Mitochondrion	An organelle that converts the energy in food into usable energy for the cell
Nucleus	An organelle that contains the cell's genetic material (DNA), which determines the nature of cell structures and substances
Ribosome	A structure that makes proteins, either free or bound to the surface of the endoplasmic reticulum
Vacuole	A membrane that stores water and other materials

Observing Fungi

Date	Description of bread (include number of colonies)	Drawing of bread
	Sample taken from	
	Observations	*Answers should include drawings and complete observations.*
	Observations	

Observing Bacteria

Date	Description of agar plate (include number of colonies)	Drawing of agar plate (include color)
	Sample taken from 1. 2. 3. 4.	1 2 3 4
	Observations 1. 2. 3. 4.	*Answers should include drawings and complete observations, including numbers of colonies.*
	Observations 1. 2. 3. 4.	1 2 3 4
	Observations 1. 2. 3. 4.	1 2 3 4

Response Sheet—Investigation 4

A student wrote the following response to the question, What are elodea plants made of?

Elodea plants are made of cells, cell walls, cytoplasm, and chloroplasts.

His friend told him that he forgot to include the levels of complexity. Improve on the first student's response, keeping in mind his friend's suggestion.

Elodea plants are made of cells. The elodea cells are made of cell structures, such as cell walls, cytoplasm, and chloroplasts. The cell structures are made of molecules, and molecules are made of atoms.

Bacterial Cell Structures and Functions

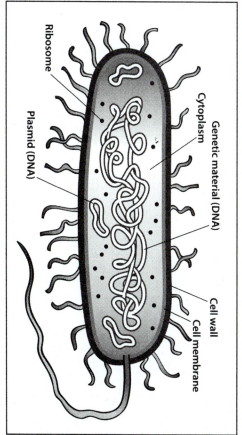

Cell structure	Function
Cell membrane	Boundary that controls what enters and leaves the cell
Cell wall	A rigid layer that supports the cell and provides shape
Cytoplasm	Internal fluid that contains the cell structures
DNA or genetic material	A molecule that determines the nature of cell structures and substances
Plasmid	Small piece of genetic material that is independent of other DNA in the cell and that can be passed to other bacteria
Ribosome	A structure that makes proteins, in the cytoplasm

Archaea

Archaea have the same cell structures as bacteria.

1. Do archaea have a nucleus? __No__

2. Do archaea have DNA and plasmids? __Yes__

3. Do archaea have cytoplasm? __Yes__

4. Do archaea have a cell wall and cell membrane? __Yes__

5. Do archaea have ribosomes? __Yes__

6. What is unique about archaea? During the class discussion, record your notes below.

   The cell wall is built differently.
   The cell membrane is different.
   The ribosomes are more like humans' than like bacteria's.
   They are found in the most extreme environments and in common environments.

Fungal Cell Structures and Functions

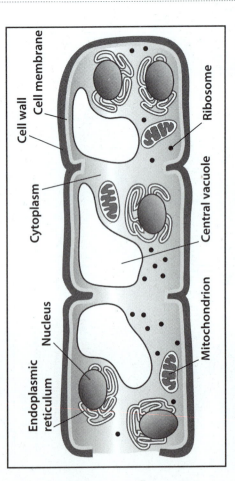

Cell structure	Function
Cell membrane	Boundary that controls what enters and leaves the cell
Cell wall	A rigid layer that supports the cell and provides shape
Cytoplasm	Internal fluid that contains the cell structures
Endoplasmic reticulum	A membranous structure that assembles proteins and parts of the cell membrane
Mitochondrion	Converts the energy in food into usable energy for the cell
Nucleus	Contains the cell's genetic material (DNA), which determines the nature of cell structures and substances
Ribosome	A structure that makes proteins, free or bound to the surface of the endoplasmic reticulum
Central vacuole	A membrane that stores water and other substances, and provides structure and support for the cell

Classification History Notes

Answer these questions in your notebook.

1. Aristotle thought that all life had ___souls___. He thought the lowest form of life was ___plants___, and the highest form of life was ___humans___ because they can reason. Linnaeus identified two kingdoms: ___animals___ and ___plants___.

2. The idea of two kingdoms remained in place until ___the late 1800s___. The ___microscope___ enabled scientists to identify single-celled organisms.

3. It wasn't until the 1950s and 1960s that scientists classified life into five kingdoms because some kinds of life did not match plants and animals. The three new kingdoms were ___Protista___, ___Fungi___, and ___Bacteria___.

4. A further classification of life depended on whether there is a nucleus in the cell. What are the two kinds of cells called? Give examples of each kind.
A prokaryote is a cell without a nucleus, like bacterial cells.
An eukaryote is a cell with a nucleus, like animal cells.

5. Another change was proposed in the 1970s, when the technology was soph[isticated]... es inside cells. Th... three domains. Th... a nucleus and orga... ok, draw how the... with the organism... lerstanding of how li... organized.

> Diagram should look like the domain diagram in FOSS Science Resources.

Celery Investigation A

	Day 0 (setup)	Day 1 (final)	Change
Water in control vial (volume)			
Water in celery vial (volume)			
Celery mass			

Part 1

1. Take the celery stalk out of the vial. Measure the amount of water... Record.
2. Reco... vial.
3. Calcu...

> The data are from Investigation 5, Part 1. Student data may vary.

s. Record.
poration)

4. How much water was lost to evaporation? ___1 mL___
5. How much water was lost in the celery vial? ___17 mL___
6. Do the amounts match? ___No___ Why or why not?

Only 1 mL was lost to evaporation. There is a 16 mL difference. It must mean that 16 mL of water are in the celery.

Part 2

7. *Predict* the current mass of the celery stalk. (Remember that for water, 1 mL = 1 g.) ___42.6 g + 16 g = 58.6 g___

8. Determine the actual mass of the celery. Record.

Celery Investigation B

Part 2 (continued)

9. Does your prediction match the actual mass of the celery?
   <u>No</u>   Record any ideas you have about your results.
   Maybe we measured wrong, but my group says we were right. The water had to have gone somewhere else. Maybe the water is lighter.

10. Calculate the *change* in mass of the celery. Record in the data table.

Part 3

11. How much of the water from the celery vial ended up in the celery? <u>Only 0.2 mL</u> How do you know?
    The mass of the celery increased only 0.2 g.

12. What do you think happened to the rest of the water that was lost from the celery vial?
    I have no idea. It went into the air? It may have spilled.

13. Determine the amount of water unaccounted for in your vial.

    17 mL   1 mL   0.2 g   15.8 mL

Part 4

14. In your notebook, describe any patterns you notice in the class celery and class data.

Leaf Observations

Tradescantia leaf (100X)

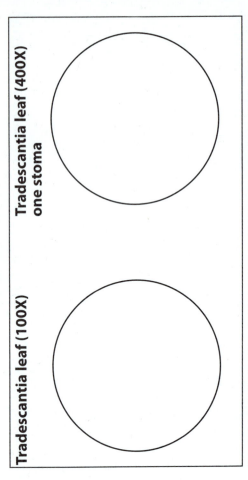

Tradescantia leaf (400X) one stoma

Crisp celery leaf (400X) (optional)

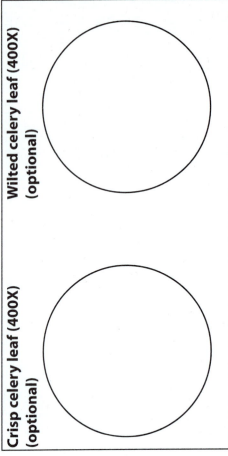

Wilted celery leaf (400X) (optional)

1. Label the guard cells and one stoma in the high-power drawing.
2. Describe the structure of a stoma.
   A stoma is a hole in a leaf. It is surrounded by two guard cells that look like lips.
3. Explain how stomata work.
   When there is a lot of water in the plant, the guard cells open the stoma and water vapor exits. If there isn't much water, the guard cells stay closed and water vapor can't exit.

Response Sheet—Investigation 5

A student noticed a plant outside that had really wilted leaves. He remarked to a friend,

Those leaves must be losing a lot of water to become so wilted. I bet that the stomata are totally open right now.

Do you agree or disagree? What would you add to the conversation?

I disagree. The stomata must have closed so that no more water will exit the leaves. There is not enough water in the plant.

Multicellular Levels of Complexity

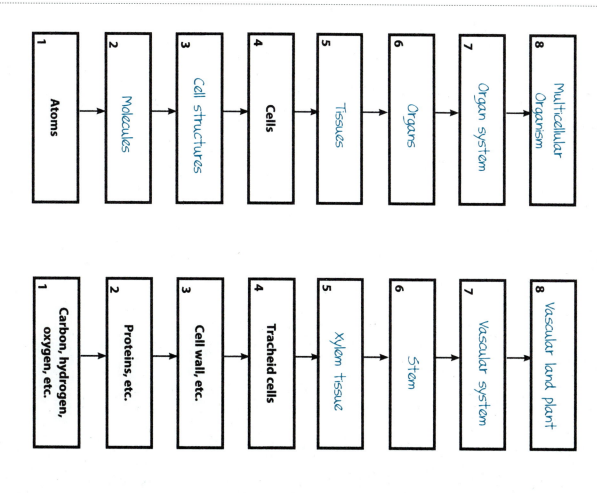

1. Atoms → 2. Molecules → 3. Cell structures → 4. Cells → 5. Tissues → 6. Organs → 7. Organ system → 8. Multicellular Organism

1. Carbon, hydrogen, oxygen, etc. → 2. Proteins, etc. → 3. Cell wall, etc. → 4. Tracheid cells → 5. Xylem tissue → 6. Stem → 7. Vascular system → 8. Vascular land plant

Germination and Growth in Different Salinities

WARNING — This set contains chemicals that may be harmful if misused. Read cautions on individual containers carefully. Not to be used by children except under adult supervision.

The kind of seed we are investigating: _____

Number of seeds in each dish: _____

1. Record the number of seeds with roots and the number of seeds with shoots in the table below.

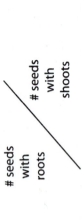
seeds with roots / # seeds with shoots

	0 spoons salt	1 spoon salt	2 spoons salt	4 spoons salt
Day 2				
Day ___				

Answers should include complete recording and observations.

2. On the final day, make your observations and comments.

0 spoons salt	1 spoon salt

2 spoons salt	4 spoons salt

Seed Dissection

Dry seed dissection: Draw and label what you observe.

Outside of seed	Inside of seed
Drawings will vary.	Drawings will vary.

Soaked seed dissection: Draw and label what you observe.

Outside of seed	Inside of seed
Drawings will vary.	Drawings will vary.

Answer these questions in your notebook.

1. How is a seed protected during dormancy? *The seed is protected during dormancy by the seed coat which is a tough, outer layer. The seed coat softens when the seed is in a suitable environment so that germination can begin.*

2. If a seed did not have cotyledons, what would happen? *If a seed did not have cotyledons, the embryo would not have food to sustain it until it started to photosynthesize.*

3. Why do you think the ability to produce seeds is an important adaptation for flowering plants? *Seeds allow plants to survive in unsuitable environments such as dry land. Each structure of the seed contributes to its survival until the embryo starts to grow.*

WARNING — This set contains chemicals that may be harmful if misused. Read cautions on individual containers carefully. Not to be used by children except under adult supervision.

Comparing Growth

Part 1: Think about the seeds you investigated.

The kind of seed we are investigating: _____

1. In which condition(s) did most of your seeds germinate? *Answers will vary, but should not include 4 spoons of salt.*

2. In which condition(s) did the fewest of your seeds germinate? *In which condition(s) did the fewest of your seeds germinate? 4 spoons of salt. May also include 2 spoons of salt.*

3. In which condition(s) do the roots and the shoots of your seeds appear the healthiest? (Compare length of roots and shoots, branching of roots, number of root hairs, greenness.) *Answers will vary, but should not include 4 spoons of salt.*

4. How does increasing the concentration of salt affect the germination and growth of your seeds? *The higher the concentration of salt, the fewer seeds germinate, and the less healthy the plants appear.*

Part 2: Compare all the seeds at each concentration of salt.

1. Which seeds (oats, wheat, barley, or corn) grew the best at 0 spoons, 1 spoon, 2 spoons, 4 spoons of salt? (Compare number of seeds germinated, healthiest looking.)

	0 spoons salt	1 spoon salt	2 spoons salt	4 spoons salt
Seed type showing most salt tolerance	Answers will vary.			

5. Which type of food crop is best suited to saline (salty) soil? *Answers will vary. Barley and/or wheat.*

6. Answer in your notebook: Is saline soil a suitable environment for germinating and growing food crops? What is your evidence? *No. The plants did not germinate as well and were not as healthy.*

Flower Dissection A

Dissection of a _____ flower

1. Look into the center of the flower. Draw a picture showing how the stamen and the pistil are arranged. Label your drawing.

   Drawings should include labeled stamens (anther and filament) and pistil (stigma and ovary).

2. Observe the end of the stamen closely. Make a close-up drawing showing the structure at the end of the stamen. Label your drawing.

   Drawings should be labeled "anther."

3. Gently push your finger into the center of the flower. Look closely at your finger with a hand lens. Describe what you see.

   Should describe pollen if present in flower.

4. If a microscope is handy, put some of the material on a slide and observe it at 100X and 400X. Draw what you see under high power. Label your drawing.

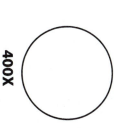

400X

Response Sheet—Investigation 6

One of your good friends was absent the day that her class discussed plant reproduction. She is trying to write a paragraph describing flowering-plant reproduction.

All I know is that baby plants come from seeds—I don't know where seeds come from.

What would you tell your friend that would help her understand how flowering plants reproduce?

A pollinator, like a bee or a hummingbird, goes to a flower to get food. It gets pollen on its body. Some of the pollen sticks to the stigma of the pistil of the next flower it visits. That is how pollination happens.

A pollen grain contains the male sex cell, the sperm. After it lands on the stigma, the pollen grain grows a tube down into the ovule in the ovary. The sperm travels down the tube to the ovule, where it joins with a female sex cell, an egg. This is fertilization.

The fertilized egg grows into an embryo, which is closed in the seed coat with the cotyledons. The embryo, the cotyledons, and the seed coat make up the seed. The seed is the living baby plant in a dormant seed. That is where seeds come from.

Flower Dissection B

5. Remove the sepals. How many are there? _____ Stick one sepal upside down on the tape near the right end.

6. Remove the petals. How many are there? _____ Stick one petal upside down on the tape next to the sepal.

7. Remove the stamens. How many are there? _____ Put all the stamens on the tape.

8. *Answers should include complete and labeled drawings and completed dissection card.*

9. Ask your teacher to cut open the ovary. Examine the inside of the ovary with your hand lens. Draw and label what you see.

   Place the pistil with the ovary cut side down on the tape next to the stamens.

10. Slide the card out from under the tape. Center the card on top of the mounted flower parts. Press down firmly to stick the card to the tape. Carefully lift up the ends of the tape and fold them to the back of the card to complete the flower mount. Label all the parts.

Pollination Syndrome A

Part 1: Observe your flower.

1. Describe the shape and color of the flower.

2. Describe any scent the flower has.

3. List any other characteristics that you think might attract pollinators.

Part 2: Use

4. Where a each oth

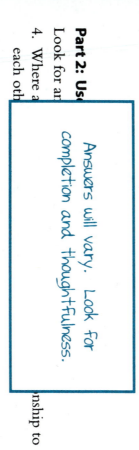

Answers will vary. Look for completion and thoughtfulness.

5. Where would a pollinator find nectar?

6. Where would a pollinator find pollen?

Pollination Syndrome B

Part 3: Think about possible pollinators.

Think about how an animal or insect pollinator might interact with your flower.

7. What characteristics might a *pollinator* have that would affect its ability to pollinate your flower?

Part 4: Use the "Flowers and Pollinators" resource.

8. Look at the tables in the resource. List your flower's characteristics

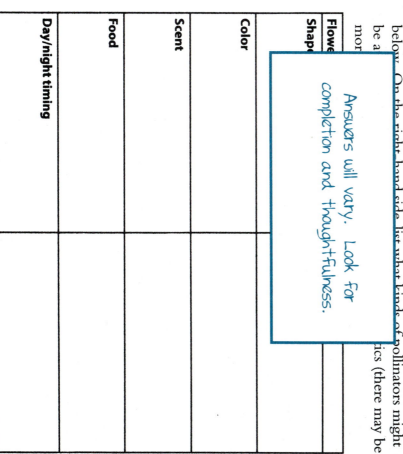

Answers will vary. Look for completion and thoughtfulness.

Flower		
Shape		
Color		
Scent		
Food		
Day/night timing		

Insect Observations A

Part 1: General insect structure
Read introduction to "Insect Structure and Function."

1. Make three drawings of the hissing cockroach: one from the top, and one from the front. Label the head, thorax, and abdomen in your drawings. Label the drawing as male or female.

2. Observe the hissing cockroach for several minutes. Describe what behaviors you observe.

Part 2: Head (eyes, antennae, and mouthparts)
Read about insect heads.

3. Label the compound eyes on one of your drawings.

4. What type of antennae do the cockroaches have? Circle below.

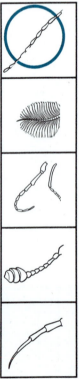

5. Hide a piece of food near the cockroach and observe its antennae. What do they do? What is the function of the antennae?

6. Use a toothpick to very carefully place a tiny bit of honey or syrup on one of the cockroach's antennae. What does the cockroach do? Why do you think this behavior is important?

7. What type of mouthparts does the cockroach have? Circle below.

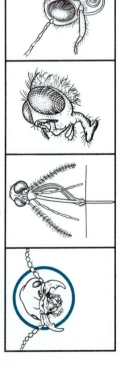

8. What kind of food do you think the cockroach eats? Put two or three different kinds of food in one of the dishes. Describe what you observe. Remove the food after your observations.

Insect Observations A

Part 1: General Insect Structure
Read Section 1 of "Insect Structure and Function."

> Drawings and answers will vary. Drawings should label head, thorax, and abdomen. Look for thoughtfulness in behavior description.

Part 2: Head (Eyes, Antennae, and Mouthparts)
Read Section 2 of "Insect Structure and Function."

5. Hide a piece of food near the cockroach and observe its antennae. What do they do? What is the function of the antennae?

   Antennae may move or wave. The function of the antennae is to sense odors and other information.

6. Use a toothpick to very carefully place a tiny bit of honey or syrup on one of the cockroach's antennae. What does the cockroach do? Why do you think this behavior is important?

   The cockroach may draw the antenna through its mouth, apparently to clean it. This helps the animal continue to sense its environment.

8. What kind of food do you think the cockroach eats? Put two or three different kinds of food in one of the dishes. Describe what you observe. Remove the food after your observations.

   Answers will vary. The cockroach might not eat at all, or might be drawn to a particular food, which the student should describe.

Insect Observations B

Part 3: Thorax (Wings and Legs)
Read about insect thoraxes.

9. Look for wings on the cockroach. Circle the type of wings it has.

10. What does this tell you about the lifestyle of the roach? The cockroach cannot fly. It lives on the ground or crawls up trees.

11. Describe how the cockroach moves. It walks, moving its six legs. The thorax seems flexible. It also grasps things and can walk up glass.

12. Circle the kind of legs the cockroach has.

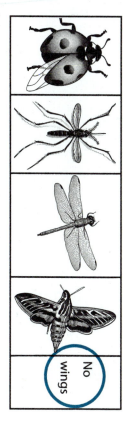

13. What part of the cockroach are the wings and legs attached to? The wings and legs are attached to the thorax.

Part 4: Abdomen
Read Section 4 of "Insect Structure and Function."

14. What is contained in the abdomen? The organs of the circulatory, digestive, and reproductive systems.

15. What are the functions of those structures? The structures of the circulatory system move nutrients and oxygen, the structures of the digestive system help the cockroach get energy from food, and the structures of the reproductive system help it reproduce.

Part 5: Behavior

16. Note the fourth segment on the abdomen of the cockroach. Can you notice the spiracles? Why do cockroaches hiss? They hiss during courtship, in defense, or in conflict with another male.

17. What questions do you have about the Madagascar hissing cockroach? List at least two. Questions will vary.

Intentionally left blank.

Structure/Behavior/Function Summary

Structure or behavior	Function(s)
Compound eye	Detects color and motion.
	Chews food.
	Senses odors, vibrations, and other things in the environment.
	Helps the cockroach crawl and run.
Hissing	Aids in defense and courtship.
Pulling antenna through mouth	Cleans antennae to better sense the environment.
Choosing dark damp places	Hides and protects the cockroach.

Comparing Systems

1. What is the transport system in each kind of organism?

   Vascular plants: __Vascular system__

   Humans: __Circulatory or cardiovascular system__

   Insects: __Open circulatory system__

2. What is the function of each system?

   Vascular plants:

   It carries water to all cells of the plant, using the xylem. It carries food from the leaves to all cells, using the phloem.

   Humans:

   It carries food and oxygen to all the cells in the body. It carries away waste from the cells.

   Insects:

   It carries food to all the cells in the body. It does not carry gases like oxygen.

3. Compare the human and insect transport systems. How are they different?

   The human system is closed and uses blood vessels and a heart. It carries food and gases. The insect system is open, which means that fluids go throughout the entire insect body. It carries only food.

4. Compare the organs of each system. How are they alike and how are they different? Make a table in your science notebook.

5. Compare the tissues of the human cardiovascular system and the insect circulatory system. How are they alike and how are they different? Make a table in your science notebook.

Comparing Systems

4.

Human transport organs		Insect transport organs
Heart	The hearts both pump blood. The dorsal blood vessel and the arteries, veins, and capillaries carry a type of blood.	A kind of heart
Arteries, veins, capillaries		Dorsal blood vessel

5.

Human transport tissues		Insect transport tissues
Cardiac muscle in heart	There is a kind of muscle tissue. There is a kind of blood.	No cardiac muscle
Smooth muscle in heart and vessels		No smooth muscle
Blood carries food and oxygen		Hemolymph carries only food

Bioblitz Summary and Reflections

How many different kinds of organisms did you predict and how many did your class actually find in the study site? Fill in the table.

Organism	Your prediction	Class count
Plants		
Fungi		
Lichens		
Animals collected		
Animal observations		

Answer these questions in your notebook.

1. Were the 5 plant, fungus, and lichen categories equal? Explain.

2. Which organisms were the most diverse at your study site? Which were the least diverse?

   Answers will vary. Look for completion and thoughtfulness.

3. Do you think the class collection accurately represents the diversity of this site? Explain.

4. What was one thing you learned from this experience?

5. What questions remain?

Intentionally left blank.

Are Viruses Living Organisms?

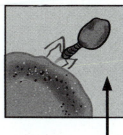
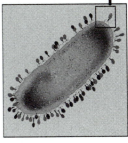

Bacteriophage T4 virus infecting an E.coli bacterium (bacterium is 200 nm long)

	Virus	Cell
Needs energy	No	✓
Needs water	No	✓
Grows	Uncertain evidence	✓ (In multicellular organisms, cells increase in number, as well.)
Reproduces	Only inside a host cell; has DNA or RNA	Asexual or sexual reproduction; DNA is genetic material.
Needs suitable environment	No	✓
Responds to environment	Uncertain	✓
Exchanges gases	No	✓
Eliminates waste	Uncertain	✓
Structure	No cell structures, but does have structure	Cell structures and organelles.
Changes over time (evolves)	Yes	✓

In your notebook, write your conclusion. Are viruses living organisms? What is your evidence? If you cannot make a decision, what other information do you still need?

Tree of Life

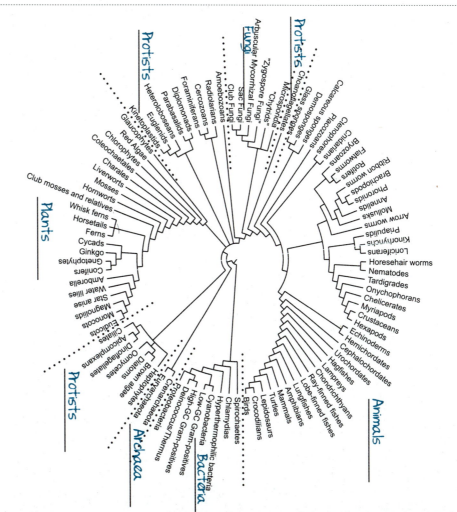

This version of the tree of life is based on data in *Life: The Science of Biology*, 9th ed., by D. Sadava, D. M. Hillis, H. C. Heller, and M. R. Berenbaum (Sunderland, MA: Sinauer Associates, 2011).